Results and Problems in Cell Differentiation

41

Series Editors
D. Richter, H. Tiedge

Brehon C. Laurent (Ed.)

Chromatin Dynamics in Cellular Function

With 14 Figures and 1 Table

 Springer

Brehon C. Laurent, PhD

Department of Oncological Sciences
Mount Sinai School of Medicine
One Gustave L. Levy Place
Box 1130
New York, NY 10029
USA

ISSN 0080-1844
ISBN-10 3-540-33685-0 Springer Berlin Heidelberg New York
ISBN-13 978-3-540-33685-3 Springer Berlin Heidelberg New York

Library of Congress Control Number: 2006925339

Springer is a part of Springer Science+Business Media

springer.com

Cover design: *Design & Production* GmbH, Heidelberg
Typesetting and Production: LE-TeX Jelonek, Schmidt & Vöckler GbR, Leipzig

Printed on acid-free paper   31/3150/YL – 5 4 3 2 1 0

Contents

Histone Ubiquitylation and the Regulation of Transcription

Histone Dynamics During Transcription:
Exchange of H2A/H2B Dimers and H3/H4 Tetramers
During Pol II Elongation

The Roles of Chromatin Remodelling Factors in Replication

Results Probl Cell Differ (41)
L. Brehon: Chromatin Dynamics in Cellular Function
DOI 10.1007/010/Published online: 17 February 2006
© Springer-Verlag Berlin Heidelberg 2006

Structure and Function of Protein Modules in Chromatin Biology

Kyoko L. Yap · Ming-Ming Zhou (✉)

Structural Biology Program, Department of Physiology and Biophysics,
Mount Sinai School of Medicine, New York University, 1425 Madison Avenue,
New York, NY 10029-6574 USA
ming-ming.zhou@mssm.edu

Abstract Chromatin-mediated gene transcription or silencing is a dynamic process in which binding of various proteins or protein complexes can displace nucleosomal histones from DNA to relieve repression or drive the gene into a highly repressed, silent state. Covalent modifications to DNA and histones associated with chromatin structural change play a crucial role in transcriptional regulation, with particular modifications on certain residues associated with a specific transcriptional outcome. In recent years a number of structural domains have been identified within chromatin-associated proteins, including DNA or RNA binding domains, protein-protein interaction domains and domains that recognize specific covalent modifications to histone tails. In this review we discuss the structural features of these protein modules and the functional roles they play in chromatin biology.

1
Introduction

Gene transcriptional regulation at the chromatin level is coordinated by a number of proteins and protein complexes that interact with nucleosomal DNA and histone proteins. The addition and removal of covalent modifications to chromatin allow for another level of transcriptional control beyond the genetic code. To attain this goal, one needs to understand the mechanisms underlying the regulation and transduction of genetic information. Growing evidence supports the view that a genome-wide *epigenetic* mechanism, imposed at the level of genomic DNA-packing histone proteins through post-translational amino acid modifications including acetylation, methylation, phosphorylation, and ubiquitination, plays a fundamental role in controlling the capacity of the genome for information storage and retrieval in response to physiological and environmental stimuli, and for inheritable changes of gene function and expression. Site- and state-specific modifications on certain amino acid residues within nucleosomal histones have been associated with a specific transcriptional outcome, e.g. gene repression or activation. Indeed, the *"histone code hypothesis"* (Strahl and Allis 2000; Turner 2002) postulates that different combinations of modifications, either in combinatorial or sequential manner, can elicit different transcriptional outcomes by recruiting proteins that recognize these modifications.

In principle, the notion of having such specific requirements to finely regulate the transcription of genes is an elegant one, and research in the last few years has focused on understanding how these epigenetic marks are created and recognized, and how such events give rise to the resulting transcriptional effects. One approach has been to examine, at the molecular and structural level, the proteins that are known to be involved in chromatin biology.

Several dozen proteins have now been implicated in chromatin remodeling, whether it be directly involved in protein modification, in the recognition of these modifications on another protein, or by virtue of its association with other known chromatin modifiers in a larger multiprotein complex (Bottomley 2004). Many of these proteins contain multiple, modular domains—conserved in sequence and/or structure—each conferring particular function(s) to the protein. These domains may occur multiple times in the same protein, and often appear in tandem with other modular domains common to other chromatin-associated proteins. Recent structural analysis of these domains at an atomic level by X-ray crystallography and NMR reveals that many of these domains exhibit folds that have already been characterized in other proteins, often of completely unrelated function. Detailed structural information about these modules has also provided clues into their function and specific roles in chromatin remodeling. Here, we review the structure and function of these conserved domains, examine the modular structural features of these proteins, and comment on the relationship between fold and function. The protein modules are grouped according to their structural folds, and their histone, nucleic acid, or chromosomal protein-binding activity.

2
Histone Lysine Acetylation Recognition by the Bromodomain

Eukaryotic genes are normally in a state of repression. Tightly wound DNA of repressed nucleosomes must be physically loosened from histones to allow transcription to occur. Modifications to residues on the protruding tails of histone octamers, such as lysine acetylation, accomplish this by reducing the charge of the tails, thereby disrupting histone-DNA, histone-protein, and histone-histone interactions (Roth et al. 2001). Acetylation on lysines is catalyzed by histone acetyltransferases (HATs), a number of which have been identified in the last several years, including Gcn5, TAFII250, CBP/p300 and p300/CBP-associated factor (PCAF). Commonly found in HATs is the bromodomain, a well-conserved protein module frequently present adjacent to PHD fingers or bromo-adjacent homology (BAH) domains (see Sect. 4). Prior to determination of its structure, bromodomains were characterized as modules of unknown function, but many proteins containing them were shown to be involved in transcription (Haynes et al. 1992; Jeanmougin et al. 1997).

Report of the first structure of a bromodomain, from PCAF (Dhalluin et al. 1999) (Fig. 1A), provided the first evidence that this protein module could interact specifically with acetylated lysines in histones. The bromodomain structure consists of a left-handed bundle of four helices (α_Z, α_A, α_B, α_C). Two loops of variable length and sequence between the helices (ZA and BC loops) form a hydrophobic pocket serving to stabilize the structure. At least five more bromodomain structures have since been solved, all exhibiting a very similar fold (Dhalluin et al. 1999; Hudson et al. 2000; Jacobson et al. 2000; Mujtaba et al. 2002, 2004; Owen et al. 2000). Biochemical and structural data of bromodomains in complex with acetyl-lysine-containing peptides (Fig. 1B) have provided important information detailing the selectivity of the bromodomain for its ligands (Kanno et al. 2004; Matangkasombut and Buratowski 2003; Zeng and Zhou 2002). The hydrophobic patch formed by the ZA and BC loops is employed in the interaction with the methyl and methylene groups of the acetylated lysine, specifically the side chains of highly conserved valine, alanine, tyrosine, and asparagine residues. Furthermore, the crystal structure of yeast Gcn5 in complex with an H4 peptide acetylated at K16 (H4-K16ac) (Owen et al. 2000) showed additional contacts between the bromodomain and residues two or three positions C-terminal to the acetylated lysine on the peptide.

The functional consequences resulting from bromodomain recognition of acetylated lysines on H3 or H4 are diverse. For example, Rsc4, a component of yeast chromatin remodeling complex RSC, contains two bromodomains

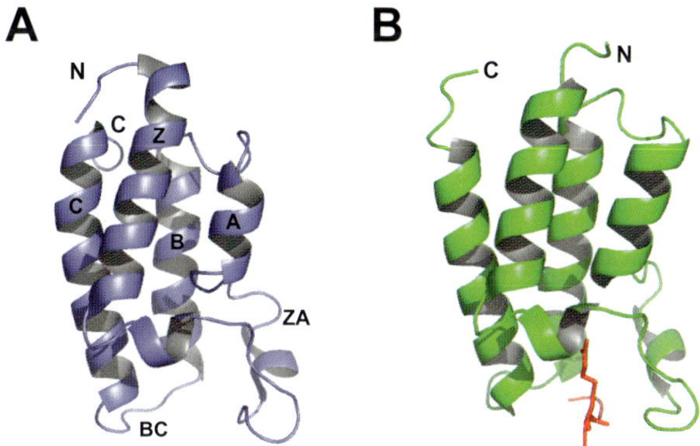

Fig. 1 Histone acetyl-lysine recognition. Acetylated lysines on the tails of histones are typically associated with transcriptionally silent chromatin, and are recognized by the α-helical bromodomain. **A** The three-dimensional structure of the bromodomain of transcriptional coactivator PCAF (PDB accession code 1N72); **B** Bromodomain of HAT Gcn5p bound to acetylated H4-K16ac (red) (1E6I)

that bind H3-K14ac, resulting in promotion of gene activation (Kasten et al. 2004). On the other hand, yeast Bdf1 competes with HDAC Sir2 for binding to acetylated H4, and prevents Sir proteins from spreading into euchromatin to silence its promoters, thereby establishing a boundary for heterochromatin (Ladurner et al. 2003). The Bdf1-H4 interaction may also compensate for an inaccessible TATA box by recruiting TFIID to its promoter to help initiate transcription (Martinez-Campa et al. 2004).

In addition to acetylated lysines in histones, bromodomains have also been demonstrated to bind acetylated lysines on non-histone proteins such as p53 (Mujtaba et al. 2004), HIV-1 Tat (Col et al. 2001; Dorr et al. 2002; Mujtaba et al. 2002), and MyoD (Polesskaya et al. 2001); bromodomains bind to un-modified histones with considerably less affinity (Ornaghi et al. 1999), but not to other modified residues. To date, the bromodomain remains the only protein module shown to bind acetylated lysine residues within a protein. Since acetyl-lysine binding occurs largely between the structurally flexible and sequence variable ZA and BC loops, the ligand binding selectivity of the bromodomain likely varies widely. Moreover, because none of the acetyl-lysine binding residues are absolutely conserved within the bromodomain family, it is possible that some bromodomains may be capable of interacting with other ligands, which remains to be verified experimentally.

3
Histone Lysine Methylation Recognition

The Royal Family Modules—The Royal family (Maurer-Stroh et al. 2003) of protein domains includes Tudor, PWWP (named for conserved proline and tryptophan residues), MBT (malignant brain tumor), Agenet and chromod-omains, characterized by an SH3-like barrel consisting of a three β-strand core. These domains frequently appear in chromatin-associated proteins, of-ten adjacent to other modular domains such as PHD and BAH domains. The plant-specific Agenet domains often occur alongside ENT domains (see Sect. 3), and although no structure is known, its sequence is sufficiently similar to be considered a distant relative of the Tudor domain (Maurer-Stroh et al. 2003). Given the known structures and functions of many of these domain-containing proteins (an association with methylation), it is believed that Royal family members have descended from a common methyl substrate-binding ancestor (Maurer-Stroh et al. 2003). For example, in addition to being found in hepatoma-derived growth factor, transcrip-tional corepressor BS69, and Wolf–Hirschhorn Syndrome protein (Stec et al. 2000), PWWP is a DNA-binding domain in DNA methyltransferases that also targets the enzyme to chromatin (Ge et al. 2004; Slater et al. 2003). The Tudor domain has been shown to bind targets containing methylated arginines (Brahms et al. 2001; Friesen et al. 2001). Intriguingly, both Tudor domains and chromodomains can interact with methylated histones. A cooperative pair of Tudor domains in the DNA damage response protein 53BP1 binds H3-K79me in an interaction necessary for targeting to double strand breaks (Huyen et al. 2004). The chromodomain of heterochromatin protein 1 (HP1) binds to H3-K9me to help establish transcriptionally silent heterochromatin (Bannister et al. 2001; Lachner et al. 2001; Nakayama et al. 2001). Drosophila Polycomb Pc binding to H3-K27me (Cao et al. 2002) functions as a dimer, recruiting two H3 tails from neighboring nucleosomes, thereby compacting the nucleosomes into a repressive state (Min et al. 2003). Finally, *Saccharomyces cerevisiae* HAT complex component Chd1 binds H3-K4me via its two chromodomains, a striking difference from the binding modes of HP1 and Pc since H3-K4me is a mark of transcriptionally competent euchromatin (Pray-Grant et al. 2005). The chromodomains of Chd1 are also required for transcriptional elongation (Simic et al. 2003).

Structurally, it is clear that these domains are related. The chromodomain contains the three-stranded β-barrel capped by a C-terminal helix (Ball et al. 1997) (Fig. 2A); the Tudor and PWWP domains have an additional helix preceding the core (Selenko et al. 2001; Qiu et al. 2002), and the latter has a C-terminal α-helical subdomain, which may be the structurally distinguishing element among PWWP domains (Nameki et al. 2005; Slater et al. 2003). MBT repeats, as found in the structures of Scm, contain an extended arm N-terminal to the fold core (Sathyamurthy et al. 2003; Wang et al. 2003); the arm of the first repeat packs against the core of the second and vice versa (Fig. 2B). HP1 contains both a chromodomain at its N-terminus and a distant relative, called a chromo shadow domain at its C-terminus (Aasland and Stewart 1995). In comparison with the chromodomain structure, the chromo shadow domain contains one helix at the N-terminus and another inserted before the C-terminal helix (Brasher et al. 2000) (Fig. 2C). Recently, the structure of a chromo barrel domain from MOF acetyltransferase was determined, illustrating another motif similar to the chromodomain, with a helix and strand at its N-terminus (Nielsen et al. 2005) (Fig. 2D).

Important functional information can be gleaned from the structural analyses of these domains in complex with their targets. Earlier research demonstrated that the SMN Tudor domain binds Sm proteins symmetrically dimethylated at arginine residues (Brahms et al. 2001; Cote and Richard 2005; Friesen et al. 2001). Comparison to the structures of the HP1 chromodomain bound to H3-K9me, -K9me$_2$, or -K9me$_3$ (i.e. mono-, di-, and tri-methylated lysine 9 of H3) revealed that the H3 binding region coincides with the Sm binding region of SMN Tudor (Jacobs and Khorasanizadeh 2002; Nielsen et al. 2002) (Fig. 2A). A very similar binding mode is displayed by the Pc chromodomain for H3-K27me binding (Fig. 2E). In both HP1 and Pc, primary interactions are with the main-chain atoms of the histone peptide (Min et al. 2003; Fischle et al. 2003); a three aromatic residue "cage" coordinates the

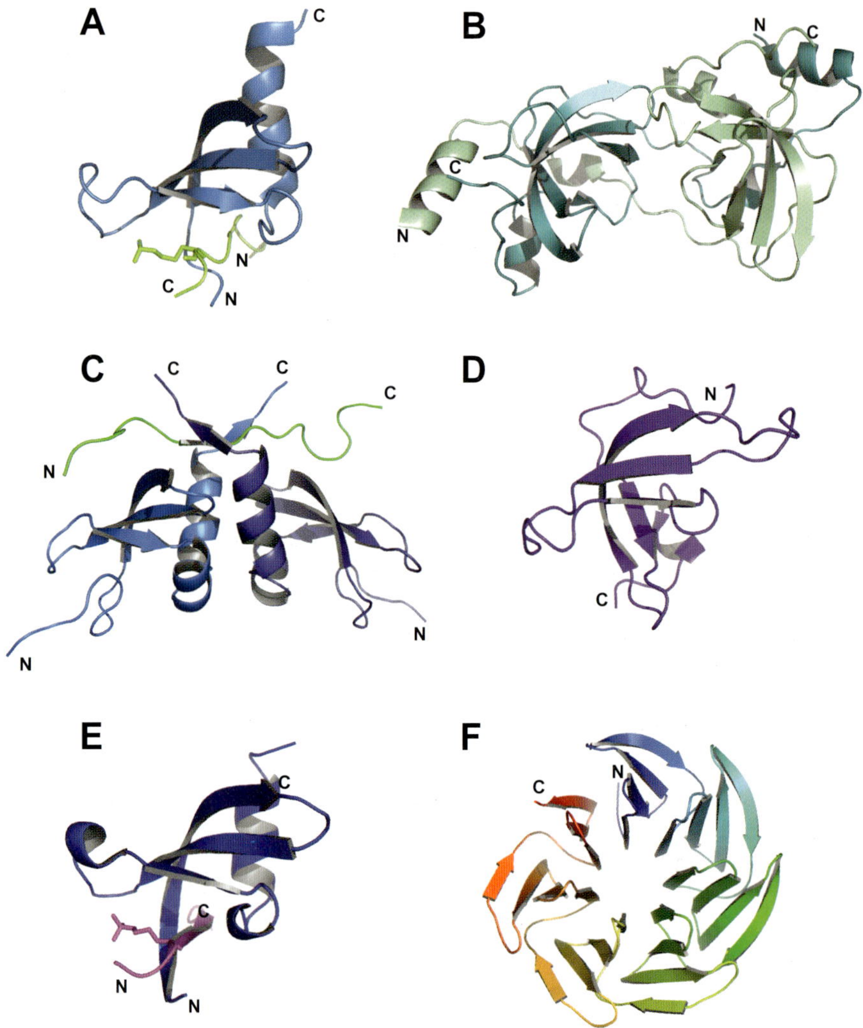

Fig. 2 Histone methyl-lysine recognition. Methylation on histone H3 and H4 lysine residues is also associated with transcriptional silencing, marks recognized by the SH3-like-barrel chromodomain. **A** Chromodomain of HP1 bound to H3-K9me (*light green*) (1GUW); **B** Two MBT repeats of Drosophila SCML2 (1OI1); **C** Chromo shadow domain homodimer of HP1 bound to Caf-1 PXVXL (*green*) (1S4Z). The binding site for Caf-1 is shared by both subunits, employing a different region of the structure compared to the HP1/H3-K9me interaction; **D** Chromo barrel domain of MOF (2BUD); **E** *Drosophila* Polycomb Pc chromodomain bound to H3-K27me (*magenta*) (1PFB). The binding region is essentially the same as that in **A**; **F** Model of WDR5 (as described in Wysocka et al. 2005) based on the structure of the C-terminal domain of Tup1 (1ERJ), which is 45% similar to WDR5. Seven WD40 repeats comprise a β-propeller fold

methyl-ammonium group of the methyl-lysine. The third of these aromatic residues are not conserved in the Chd family, possibly influencing the manner of its binding to H3-K4me (Pray-Grant et al. 2005). Although it does not interact with histones, the chromo shadow domain of HP1 binds a PXVXL motif of nucleosome assembly/DNA repair factor Caf-1 (Thiru et al. 2004) (Fig. 2C) and a RLVPL sequence adjacent to ENT of EMSY (Ekblad et al. 2005). The chromo shadow domain is a dimer in the former interaction, and the peptide forms a β-sheet with residues from the C-terminus of each HP1 monomer. On the basis of biophysical, NMR, and modeling studies, it is believed that an HP1β dimer interacts to a dimer of EMSY in a similar manner, using the same binding interface, although a difference in target sequence suggests some minor differences in coordination (Ekblad et al. 2005). Dimerization of HP1 via its chromo shadow domain may be vital for its localization to heterochromatin (Ekblad et al. 2005), and it may confer additional repressive ability in a manner similar to Pc.

The chromo barrel domain of MOF HAT was previously characterized as an RNA-interaction module, with the binding of roX2 RNA *in vivo* that is essential for MOF association with the male X chromosome (Akhtar et al. 2000). However, RNA electrophoretic mobility shift assay studies on the isolated domain demonstrated that while it may be necessary, it is not sufficient for MOF interaction with RNA (Nielsen et al. 2005).

The WD40 Repeat—Another chromatin module implicated in histone methyl-lysine binding is the WD40 repeat, a motif usually starting with Gly-His residues and ending with Trp-Asp —hence its name—with a core of about 28 residues (Smith et al. 1999). It is found in many chromatin-associated proteins, such as EED (an essential part of the PRC2 complex), RbAp46 and RbAp48 (found in chromatin-associated complexes), and WDR5 (associated with MLL complexes). The WD40 motif is one blade of a β-propeller fold, consisting of a small four-stranded β-sheet. The β-propeller is a stable and closed circular structure and can coordinate sequential or simultaneous interactions with several other proteins (Hennig et al. 2005). Multicopy suppressors of *ira1*-like (MSIL) proteins are a large sub-group of the WD40 family, containing seven repeats, and may interact directly with histone H4 prior to chromatin assembly (Vermaak et al. 1999; Verreault et al. 1998). It has thus been proposed that MSIL proteins may maintain heritable epigenetic patterns during nucleosome assembly and exchange (Hennig et al. 2005). WDR5 is a highly conserved WD40-repeat protein (Fig. 2F) and a component of several H3-K4 methyltransferase complexes, and has been shown to bind to H3-K4me$_2$, enabling (and possibly promoting) di- to trimethyl conversion. This interaction can enhance transcription, as shown when tethered to a reporter gene (Wysocka et al. 2005). Furthermore, its function is essential for the development of *Xenopus laevis*, underscoring the importance of WDR5-mediated H3-K4 methylation.

4
Chromosomal DNA/Histone Binding

The homeodomain—The homeodomain is a highly conserved domain found in a diverse range of species, particularly in eukaryotic transcription factors (Banerjee-Basu et al. 1999). Its structure is a globular, compact bundle of three to five helices, at least one of which is responsible for DNA binding specificity. At least three distinct domains found in chromatin-associated proteins—SANT, SLIDE, and ENT—are structurally related to the homeodomain. SANT (Swi3, Ada2, N-CoR, TFIIB B) domains are three-helical bundles resembling the DNA-binding domain of Myb-related proteins; however, residues that are important for coordinating DNA in the Myb structure are not conserved in SANT domains and the overall surface is more acidic. Instead, SANT domains have been identified in a number of chromatin-remodeling complexes, including SWI/SNF, RSC, and Gcn5 HAT complexes (Boyer et al. 2004). Mutational and deletion analysis indicate that the SANT domain in these complexes is indispensable for activity, illustrated by a decreased ability to interact with non-acetylated histone H3 tails (Boyer et al. 2002) and an inability to interact with partner co-repressor proteins (Yu et al. 2003). A SANT domain in the SMRT co-repressor also binds to non-acetylated, but not tetra-acetylated, histone H4 (Yu et al. 2003). In the chromatin remodeling ATPase ISWI, the SANT domain (Fig. 3A) is followed by a second, four-helical bundle SANT-like domain of unique sequence, termed the SANT-like ISWI domain or SLIDE (Grune et al. 2003) (Fig. 3B). Intriguingly, SLIDE contains many of the conserved DNA interacting residues found in Myb, has a generally basic surface and binds synthetic, Holliday-junction DNA. These features suggest cooperative interactions of SANT and SLIDE with the nucleosome, in which SLIDE could bind nucleosomal DNA and the SANT domain could interact with the histones. A third ISWI domain, called the HAND domain, may also be involved in histone binding (see below).

ENT (EMSY N-Terminal) domains have been identified primarily in plant species, where they are frequently accompanied by Agenet domains. As Agenet domains are part of the larger Royal family of domains often associated with chromatin structure regulation, it has been speculated that ENT domains also play some role in chromatin remodeling, although a specific function has yet to be identified. Recently, EMSY was found to dimerize via ENT, providing a platform for dimeric HP1β binding to a region C-terminal to ENT (Ekblad et al. 2005). The only known eukaryotic ENT domain is found at the N-terminus of EMSY, a BRCA2-associated protein amplified in some sporadic breast and ovarian cancers. Its positively charged surface suggests it has the potential to bind DNA directly, although this function has not yet been shown. The EMSY ENT domain forms a homodimer, which may enable EMSY to interact with two binding partners simultaneously (Chavali et al. 2005). Structurally, it is a five-helical bundle that resembles both SAND and

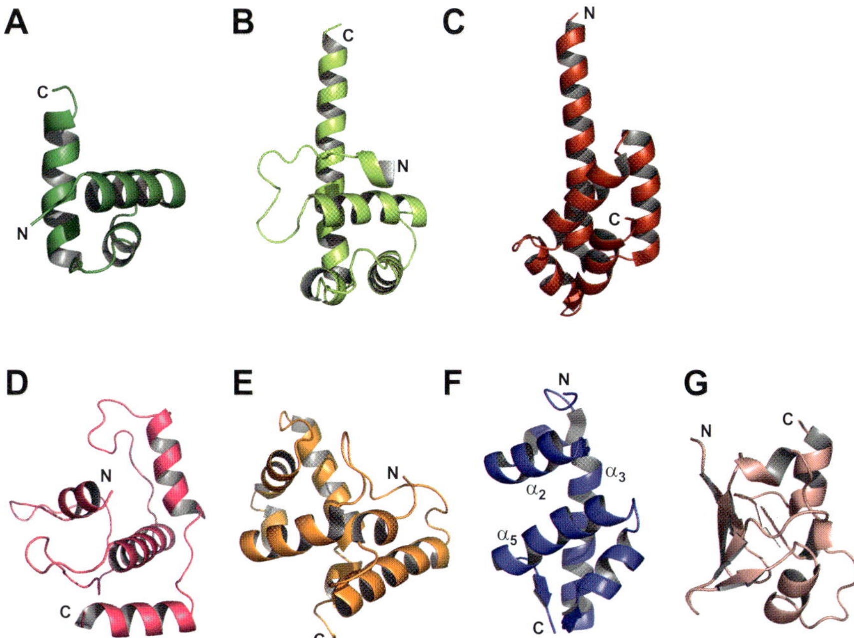

Fig. 3 Chromosomal DNA or histone recognition. Most of these domains are all-helical, with the exception of the SAND domain, which is helical on one face and β-strand on the other. **A** The SANT domain; and **B** SLIDE of ISWI (1OFC). These two helical bundles, found in tandem in ISWI, may cooperate to bind the nucleosome, with the SANT domain interacting with the histone, and SLIDE interacting with the associated DNA; **C** ENT of EMSY (1UZ3), the dimerization domain that enables EMSY interaction with HP1β; **D** The N-terminal HAND of ISWI (1OFC); **E** ARID from SWI/SNF complex protein p270 (1RYU), a non-specific DNA binding domain; **F** ADA2α SWIRM (2AQF), which binds to dinucleosomal DNA on a surface formed by α_2, α_3, α_5 and proximal interhelical loops; **G** SAND domain of Sp100b (1H5P), which interacts with DNA via the basic, helical half of its surface

SLIDE domains, despite a different ordering of helices in forming the overall structure (Fig. 3C).

The HAND Domain—The HAND domain is so named because of the positions of its four helices—three resembling an open hand and the fourth as a thumb in its palm (Fig. 3D). It has thus far only been identified in the nucleosome remodeling ATPase ISWI enzyme, has no sequence similarity to any other known protein or domain and has no structural homolog to any other solved structure in a DALI search (Grune et al. 2003). In ISWI, the HAND domain is sequentially the first in the C-terminal half of the structure, followed by a SANT domain, a spacer helix, and SLIDE. It has no known function, but because of its close proximity to the SANT domain and the presence of interacting residues that form a common hydrophobic cluster, it is believed that the HAND and SANT domains together should be considered a single, cooperative unit.

The ARID motif—The A-T rich interaction domain (ARID) motif is a DNA binding, five or six-helical bundle that is found in many proteins involved in development, cell differentiation, and chromatin remodeling (Kortschak et al. 2000; Patsialou et al. 2005; Wilsker et al. 2005). Although initially identified as binding specifically to A-T rich sequences, many ARIDs bind DNA non-specifically. The chromatin remodeling complexes Brahma and SWI/SNF include ARID-containing proteins (such as Osa, SWI1 in yeast and p270/SMARCF1 and ARID1B in human) but the DNA-binding function of these proteins does not appear to be responsible for the complex's ability to bind DNA (Kortschak et al. 2000). The ARID of p270, which contains one additional helix at the N-terminal end of the motif (Kim et al. 2004) (Fig. 3E), binds the minor and major groove of DNA via the flexible linkers between helices 1 and 2, and helices 4 and 5 of the domain, respectively (Iwahara et al. 2002; Zhu et al. 2001). The linkers are considerably longer than that in the common helix-turn-helix DNA binding motif. p270 ARID has been shown to bind DNA without sequence specificity, and therefore is not believed to recruit SWI/SNF complexes to specific promoter sequences (Wilsker et al. 2004). DNA binding via yeast SWI1 ARID is also believed to be non-specific and weaker than that of p270 (Wilsker et al. 2004).

The SWIRM domain—The SWIRM domain is found in a number of chromatin remodeling or modification-related proteins, including Swi3, Rsc8, Moira, and LSD1, and thus may be involved in histone lysine demethylation (Shi et al. 2004), chromosome synapsis (Aravind and Koonin 1998) and replication fork protection and strand synthesis (Noguchi et al. 2003). Originally predicted to be a protein–protein interaction domain (Aravind and Iyer 2002), the SWIRM domain of transcriptional activator ADA2α was recently characterized as a DNA binding domain that is capable of binding to both double-stranded free and chromosomal DNA (Qian et al. 2005). Moreover, ADA2α SWIRM co-localizes with lysine-acetylated histone H3 in the nucleus, and has a direct role in ATP-dependent, ACF-mediated chromatin remodeling by increasing accessibility of histone H1-bound dinucleosomal DNA.

The structure of ADA2α SWIRM is a 5-helix bundle consisting of two helix-turn-helix motifs connected by a central long helix, reminiscent of the histone fold (Fig. 3F) (Qian et al. 2005). The three helices ($\alpha_3 - \alpha_4 - \alpha_5$) and small capping β-sheet (consisting of strands before α_4 and after α_5) that comprise the winged-helix DNA binding motif found in transcriptional factors ETS1 (Kodandapani et al. 1996) and E2F4 (Zheng et al. 1999) and linker histone H5 (Ramakrishnan et al. 1993) constitute a hydrophobic core characteristic of this motif. Residues that may be involved in the interaction with DNA locate to α_2, α_5 and the loops between α_1, α_2 and α_3, and primarily occupy one face of the protein surface. On the basis of this structure and the sequence alignment of known SWIRM domains, it was proposed that the basic winged-helix motif would be conserved, but sequence and structure variance at the N-terminus indicate three possible sub-groups, with ADA2α in one, transcriptional activators Swi3, Rsc8, and Moira in another sub-group, and the histone lysine demethylase LSD1 (also known as BHC110 or AOF2) in the third.

The SAND Domain—Named after Sp100, AIRE-1, NucP41/75, and DEAF-1, the SAND domain often occurs in modular proteins alongside bromodomains, PHD fingers, and MYND domains. As no SAND domain has been found in yeast, it is possible that it is restricted to animal species (Gibson et al. 1998). The structural core consists of four short helices packed against a half-β-barrel (of five or six strands), somewhat resembling the SH3-like fold (Bottomley et al. 2001; Surdo et al. 2003). The SAND domain has primarily been characterized as a DNA-binding domain. In nuclear DEAF-1 related (NUDR) protein, the SAND domain mediates DNA binding with an interacting surface involving the positively charged surface of the helical half of the fold and a conserved KDWK motif (Bottomley et al. 2001) (Fig. 3G). The SAND domain of glucocorticoid modulatory element binding proteins 1 and 2 (GMEB1, GMEB2) helps coordinate zinc in addition to binding DNA, but the role of zinc binding has not been determined. It also employs the KDWK motif on the α-helical face for DNA binding (Surdo et al. 2003). The glucocorticoid receptor binding site on the protein overlaps with the SAND domain, and binding may involve the β-sheet surface.

5
Chromosomal Protein–Protein Interactions

The BAH Domain—The Bromo-adjacent homology (BAH) domain has no defined function but it appears alongside other chromatin modular domains and in proteins involved in transcriptional silencing, DNA replication, and DNA methylation (Callebaut et al. 1999). The hexa-bromodomain protein polybromo is a key component of a SWI/SNF related ATP-dependent chromatin remodeling complex PBAF (SWI/SNFβ), which is related to the yeast RSC complex. RSC1 and RSC2 each contain one BAH domain, while both DNA methyltransferase DNMT1 and polybromo contain a pair of tandem BAH domains. Aside from what could be inferred from the general functions of BAH-containing proteins, little more could be deduced about the specific role of the BAH domain, except speculation that it is involved in protein–protein interactions (Goodwin and Nicolas 2001).

However, recent work, including structural studies, has started to shed light on this motif. The conserved domain is approximately 133 residues; the structure of the proximal BAH domain of polybromo is a distorted β-barrel (Fig. 4A). On the basis of sequence alignment, polybromo, RSC, and transcription factor Ash1 BAH domains are more related to each other than to the BAH domains of origin recognition complex 1 protein (Orc1p) and DNMT1

(Hou et al. 2005; Hsu et al. 2005; Oliver et al. 2005). Recent structures of Orc1p N-terminal domain confirms this, with the BAH domain structure juxtaposed with an additional helical "H-domain" integrated into the middle of the sequence, physically residing outside one half of the barrel. This non-conserved helical domain is responsible for binding Sir1p (the interacting domain of which is a relative of the chromo- or chromo barrel domain) (Zhang et al. 2002). Interestingly, the Drosophila homolog of Orc1p binds HP1 at this region (Pak et al. 1997).

The PHD Finger—The plant homeodomain (PHD) finger is a small (50–80 amino acids), zinc-binding, cysteine-rich motif typically occurring alongside other modular domains, and is generally considered a protein interaction domain (Aasland et al. 1998). The structure of the zinc-bound PHD finger from human Mi2-β chromodomain-helicase-DNA-binding protein 4 (CHD4) illustrates a fold with little regular secondary structure, except for a small β-sheet in the middle of the domain (Kwan et al. 2003) (Fig. 4B). It is found in proteins involved in chromatin-mediated transcriptional regulation and X inactivation (Aasland et al. 1998) and has been directly implicated in E3 ubiquitin ligase activity (Lu et al. 2002), transcriptional repression (Aapola et al. 2002; Shamay et al. 2002), acetyltransferase activity (Kalkhoven et al. 2002) and phosphoinositide binding (Gozani et al. 2003). Mutations in the PHD finger of CBP as found in Rubinstein–Taybi syndrome result in a loss of CBP acetyltransferase (AT) function (Kalkhoven et al. 2003). Interestingly, while the PHD finger is necessary for p300 AT activity, including CBP autoacetylation and *in vitro* acetylation of core histones (Kalkhoven et al. 2002) it is dispensable for AT activity of p300, illustrating a fundamental difference from these highly similar enzymes (Bordoli et al. 2001).

The PHD finger may directly interact with histones, but it appears it may not bind the N-terminal tails as is commonly seen among other chromatin remodeling domains. ACF1 and ISWI (which contains HAND, SANT, and SLIDE domains as described above) are components of ACF and CHRAC complexes, which are involved in nucleosomal array assembly (Eberharter et al. 2001; Ito et al. 1999; Poot et al. 2000). Deletion of the two C-terminal PHD fingers of ACF1, or disruption of their zinc binding ability, dramatically reduces the energy efficiency of ISWI activity, including the ability to slide the mononucleosome along a DNA fragment (Eberharter et al. 2004). As zinc is necessary for proper folding of the PHD finger (Capili et al. 2001; Pascual et al. 2000), this defect is a result of a dramatic structural compromise. The ability of the ACF1 PHD fingers to mediate nucleosomal mobilization is due to its interaction with the four histones. The PHD fingers bind H2A, H2B, H3, and H4 histones equally, and as strongly to tail-less histones, indicating that binding does not occur at the flexible, non-conserved tails, but at the common central histone region.

Studies have shown that the frequently occurring tandem PHD finger-bromodomain behaves as a cooperative unit. For example, in HAT p300 the

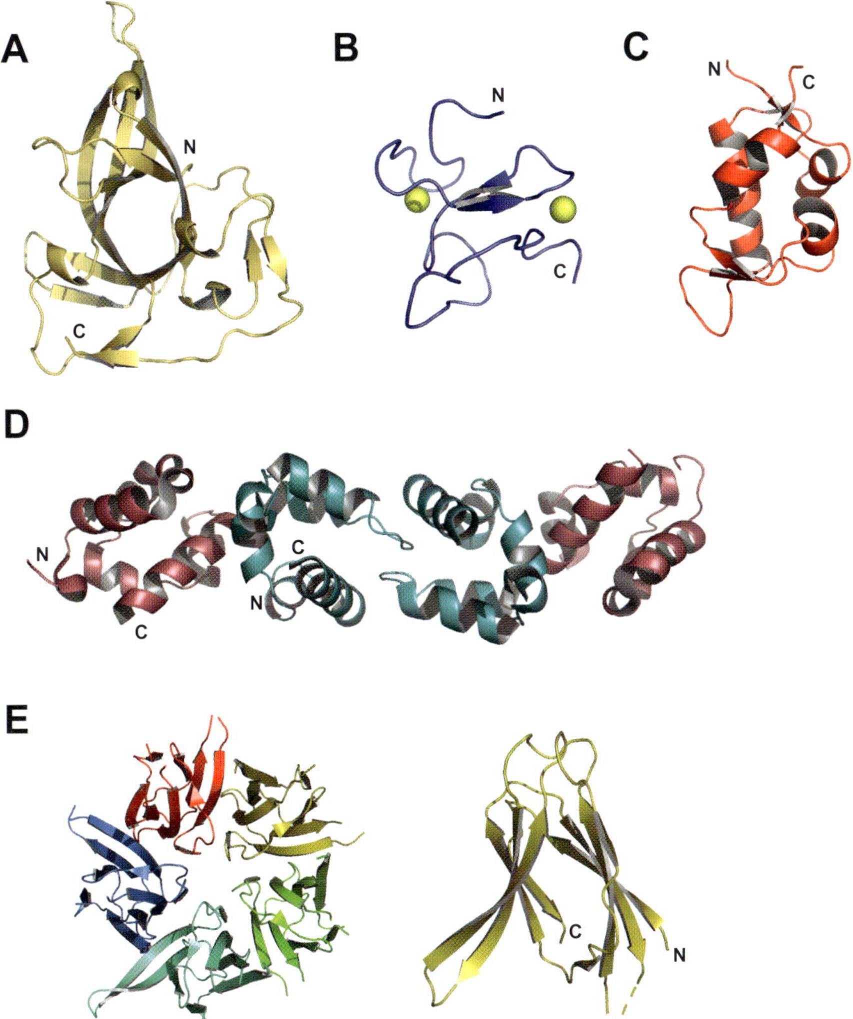

Fig. 4 Chromosomal protein–protein interactions. **A** Proximal BAH domain of polybromo (1W4S), a distorted SH3-like barrel. The BAH domain of Orc1 (not shown) is involved in interactions with Sir1p, and HP1; **B** PHD finger from CHD4 (1MM2), with zinc ions shown in *yellow*. The PHD finger likely functions in cooperation with an additional structural domain; **C** BAF60b SWIB (1UHR); **D** Copolymer of Ph/Scm SAM domains (Ph, teal; Scm, pink) (1PK1), which inhibits nucleosome remodeling by SWI/SNF; **E** Xenopus nucleoplasmin core, pentamer and monomer (1K5J). Two pentamers, forming a decamer, is believed to help dock histone H2A/H2B dimers, onto which an H3/H4 tetramer would assemble into a histone octamer

PHD finger-bromodomain binds highly acetylated nucleosomes, and both domains are required for the interaction (Ragvin et al. 2004). Additionally, the PHD finger-bromodomain of chromatin remodeling complex NoRC interacts with H4-K16ac, enabling heterochromatin formation and silencing mammalian rRNA genes (Zhou and Grummt 2005). In fact, an adjacent domain may even quench the apparent function of the isolated PHD finger. In plant homeodomain transcription factors, where PHD fingers were first identified, the leucine zipper often precedes the PHD finger to form a ZIP/PHD motif, and it was reported that the isolated PHD finger's transcriptional activation ability is masked when it is a part of the ZIP/PHD motif (Halbach et al. 2000).

The SWIB domain—Besides occurring in the p53 antagonist MDM2 in the p53-binding region (Bennett-Lovsey et al. 2002), the SWIB domain is also found in SWI/SNF related regulator of chromatin BRG1-associated factor 60a (BAF60b). BAF60b is a muscle-specific member of the BAF60 family found in mammalian SWI/SNF complexes such as SWIB (SWI complex B). SWIB (as an isolated protein) may have been acquired by chlamydia bacteria from a mammalian host, playing a role in chromatin condensation–decondensation (Bennett-Lovsey et al. 2002). The structure of the mouse BAF60b SWIB domain is a four-helical open bundle capped on either end by a small β-sheet (Fig. 4C).

The SAM domain—The Sterile alpha motif (SAM) domain (also known as Pointed or Helix-Loop-Helix domains) is found in well over a thousand proteins of vastly differing functions. The number of SAM domains occurring in a genome is more or less correlated with the complexity of the organism (Qiao and Bowie 2005). Primarily known as a protein interaction domain, it is often found as a homo- or hetero-dimer with SAM domains of other proteins. It can also form polymers, as found in the TEL transcriptional repressor, ETS proteins and *Drosophila* Polycomb group chromatin remodeling proteins (Kim et al. 2002; Qiao and Bowie 2005). The overall structure of the SAM domain is of two orthogonal α-hairpins containing four or five helices. The crystal structures of TEL-SAM and a Polycomb polyhomeotic (Ph)/Sex-comb-on-midleg (Scm) heterodimer (Kim et al. 2005) (Fig. 4D) are head-to-tail polymers with remarkable structural similarity, given their rather disparate sequences. Mutations disrupting polymerization of TEL prevent transcriptional repression (Wood et al. 2003), illustrating the importance of polymerization to activity of the protein and suggesting its relevance to Polycomb-mediated repression. The polymeric interface between SAM domains involves the so-called mid-loop (ML) and end-helix (EH) surfaces, composed of residues from $\alpha_2 - \alpha_3 - \alpha_4$ and α_5, respectively. Polycomb group proteins inhibit nucleosome remodeling by the SWI/SNF complex, and formation of the co-polymeric structure supports a model in which Ph forms a polymer in the vicinity of a Polycomb response element, with Scm complexes extending the polymer, thereby enhancing spreading by Polycomb complexes on chromatin (Kim et al. 2005). In addition to its more characterized function as a protein–protein interaction domain, studies on SAM from translational repressor Smaug revealed that the SAM domain is also an RNA-interaction motif. The positively charged surface proposed to be involved in RNA binding is conserved among Smaug homologues, but not among other SAM domains (Green et al. 2003). Very recently, a number of RNA-bound SAM domain structures of one such homolog, the yeast post-transcriptional regulator Vts1p, have been reported (Edwards et al. 2005; Aviv et al. 2006; Johnson and Donaldson 2006; Oberstrass et al. 2006). The high affinity interaction involves a guanosine base within the RNA stem loop and several hydrophobic and basic residues within helices $\alpha1$ and $\alpha2$ and helix $\alpha5$ of the SAM domain. Recognition of RNA by Vts1p is described as both sequence- and shape-dependent, based on Vts1p's ability to bind certain loops of differing sequences (Aviv et al. 2006; Oberstrass et al. 2006).

The Nucleoplasmin-like Core Domain—Nucleoplasmin (Np) is a phosphoprotein involved in chromatin decondensation and nucleosome assembly (Ito et al. 1996). It functions as a histone chaperone or possibly as part of a histone storage complex (Arnan et al. 2003). Np binds H2A/H2B dimers to assemble histone octamers, potentially into a decamer (Dutta et al. 2001). In the solved structure of the Np core (N-terminal) domain, one monomer consists of an eight-stranded β-barrel, stably folding into a pentamer (Fig. 4E), and associating again to form a decamer. It was hypothesized that the decamer might function as a docking ring for pairs of H2A/H2B dimers, each of which would then recruit a tetramer of H3/H4. Computational docking confirmed the compatibility of the Np decamer surface with that of the histone octamer. The structure of the core domain of *Drosophila* Np-like protein (NLP-core) was also solved and strongly resembles that of Np-core, except for a β-hairpin that is extended in the NP-core decamer structure and is more compact in the NLP-core structure (Namboodiri et al. 2003). NLP may also mediate chromatin decondensation in sperm, and since its levels are highest in the early embryonic stage, NLP may be a developmentally regulated histone chaperone.

6
Discussion

The overview of the structures and functions of conserved protein modules found in chromatin associated proteins illustrates the diversity in epigenetic signals that regulate transcription, the structural scaffolds that recognize these signals, and the downstream cellular responses. Some domains are involved in the direct readout of the histone code by recognizing specifically modified amino acids in histones (including the bromodomain and the chromodomain) and some interact with other chromosomal DNA or proteins (e.g.

the SWIRM domain and the PHD finger). Many of these domains are found in proteins related to other aspects of transcriptional regulation or in proteins of seemingly completely unrelated function. It is intriguing that these conserved folds can be found in such different contexts, underscoring the idea that there probably are a finite number of folds in the protein fold universe, imposed by structural and surface-associated energetic limitations.

It is therefore not difficult to see that the structures of these protein folds are more conserved than their sequences. Despite the wealth of genomic sequence information and increasingly powerful structure prediction algorithms available today, newly discovered protein sequences can still evade domain prediction if the sequence similarity to other known proteins is poor. Once the structure is known, however, sequence differences found among a fold family can encode valuable information, determining different interaction partners and ligands and the specificity of these interactions. Notably, we observe in some of these protein domains with the same fold that the most variable regions, in both sequence and structure, often are directly involved in interaction with their binding targets.

In the most general sense, it follows that these modular domains are more likely to retain a common function than a common sequence motif, and that function is tied more strongly to structural fold than to sequence. A protein known to have the same function in humans as another protein found in yeast is far more likely to have a similar fold than similar sequence. One needs to exert caution, however, in predicting the function from a structure without other biological evidence, given the propensity of some common folds to be involved in a variety of completely different functions.

That many of the domains found in chromatin biology have multiple functions complicates any kind of classification or simple correlation of fold with function. An overall trend that may be apparent from the discussion above is the predominance of DNA-binding modules that are primarily α-helical folds (the homeodomain, ARID and winged-helix/histone fold/SWIRM domains). One of two RNA-binding modules is a primarily β-strand fold (chromo barrel; Smaug SAM domain is all-helical). While this is by no means a steadfast rule (an expansion to other non-chromatin remodeling domains might better illustrate this pattern), the general explanation for this trend is that α-helical structures are generally flexible, thereby accommodating the rigidity of DNA (Qian et al. 2005; Zheng et al. 1999); conversely, β-strand structures are more rigid which allows it to bind to the more flexible RNA. On the other hand, protein interaction domains accommodate more targets with varying structural attributes, and thus are likely to employ more diverse motifs.

The fraction of multiple-domain proteins found in eukaryotic species is estimated to be 65%, and 8% of multi-cellular proteomes are domain repeats (Ekman et al. 2005). Thus, it is hardly surprising to find that several chromatin-associated domains are found in multiples and alongside other functional domains within proteins. Protein interaction domains will recruit other proteins for modulation or additional interaction; domains may work cooperatively to provide additional stability (e.g. the PHD finger/bromodomain), a shared catalytic site or interaction surface (e.g. potential simultaneous nucleosome binding by the SANT/SLIDE domains), or other enhancement of function. Repeats of motifs can form a domain in itself (WD40), provide additional specificity (such as the six bromodomains of polybromo, which may target multiple modified ligands) or cooperatively function (both PHD fingers of ACF1 bind more strongly to the core histones than individually). Multiple combinations of these domains can allow a single protein to exert several actions sequentially or simultaneously, and these permutations help explain why one fold could appear to have so many functions.

7
Future Directions

The understanding of any biological system requires a thorough, multidisciplinary approach, and should involve analysis at the molecular level of the proteins involved. As shown here, atomic resolution structures determined by NMR and X-ray crystallography provide valuable insight into protein function—by illustrating interaction surfaces for other proteins or nucleic acids, explaining functional mutations, and even pointing to unexpected roles in chromatin biology. With this approach, biologists have made progress in understanding many aspects of chromatin-mediated gene transcription or silencing.

However, it is likely that there are still many chromatin remodeling protein modules as yet unknown and undiscovered. For example, a number of components of the larger chromatin remodeling complexes remain uncharacterized both biochemically and structurally. As has been described above, we have, for several domains, a determined three-dimensional structure and only a vague understanding of its function. Furthermore, assuming the existence of a histone code, a multitude of combinations of modifications are possible, and we still have little information on how multiple modifications can be recognized—and whether this will require a "new fold" or an "old fold". Clearly, establishing the determinants for chromatin-directed transcriptional regulation has only begun.

8
Concluding Remarks

With the completion of the human genome project we now have an understanding of the sequences and organization of genes that control every biological process in the cell. However, we still know little about the function of many of the proteins encoded by these genes. Thus, the subsequent focus on describing the proteome has justifiably shifted toward a more structure-based approach. This concept of characterizing the proteome with structural biology tools has been demonstrated in established fields such as signal transduction, but not fully in relatively newer research areas such as chromatin signaling and RNA interference. This review has illustrated that such an approach can be beneficial to untangling the complexities of chromatin biology and regulation.

Acknowledgements This work was supported by a fellowship from the Terry Fox Foundation/National Cancer Institute of Canada (to K.L.Y.) and by grants from the National Institutes of Health (to M.-M.Z.).

References

Aapola U, Liiv I, Peterson P (2002) Imprinting regulator DNMT3L is a transcriptional repressor associated with histone deacetylase activity. Nucleic Acids Res 30(16):3602–3608

Aasland R, Gibson TJ, Stewart AF (1995) The PHD finger: implications for chromatin-mediated transcriptional regulation. Trends Biochem Sci 20(2):56–59

Aasland R, Stewart AF (1995) The chromo shadow domain, a second chromo domain in heterochromatin-binding protein 1, HP1. Nucleic Acids Res 23(16):3168–3173

Akhtar A, Zink D, Becker PB (2000) Chromodomains are protein-RNA interaction modules. Nature 407(6802):405–409

Aravind L, Iyer LM (2002) The SWIRM domain: a conserved module found in chromosomal proteins points to novel chromatin-modifying activities. Genome Biol 3(8):RESEARCH0039

Aravind L, Koonin EV (1998) The HORMA domain: a common structural denominator in mitotic checkpoints, chromosome synapsis and DNA repair. Trends Biochem Sci 23(8):284–286

Arnan C et al. (2003) Interaction of nucleoplasmin with core histones. J Biol Chem 278(33):31319–31324

Aviv T, et al. (2006) Sequence-specific recognition of RNA hairpins by the SAM domain of Vts1p. Nat Struct Mol Biol 13(2):168–176.

Ball LJ et al. (1997) Structure of the chromatin binding (chromo) domain from mouse modifier protein 1. Embo J 16(9):2473–2481

Banerjee-Basu S et al. (1999) The Homeodomain Resource: sequences, structures and genomic information. Nucleic Acids Res 27(1):336–337

Bannister AJ et al. (2001) Selective recognition of methylated lysine 9 on histone H3 by the HP1 chromo domain. Nature 410(6824):120–124

Bennett-Lovsey R et al. (2002) The SWIB and the MDM2 domains are homologous and share a common fold. Bioinformatics 18(4):626–630

Bordoli L et al. (2001) Functional analysis of the p300 acetyltransferase domain: the PHD finger of p300 but not of CBP is dispensable for enzymatic activity. Nucleic Acids Res 29(21):4462–4471

Bottomley MJ et al. (2001) The SAND domain structure defines a novel DNA-binding fold in transcriptional regulation. Nat Struct Biol 8(7):626–633

Bottomley MJ (2004) Structures of protein domains that create or recognize histone modifications. EMBORep 5(5):464–469

Boyer LA et al. (2002) Essential role for the SANT domain in the functioning of multiple chromatin remodeling enzymes. Mol Cell 10(4):935–942

Boyer LA, Latek RR, Peterson CL (2004) The SANT domain: a unique histone-tail-binding module? Nat Rev Mol Cell Biol 5(2):158–163

Brahms H et al. (2001) Symmetrical dimethylation of arginine residues in spliceosomal Sm protein B/B′ and the Sm-like protein LSm4, and their interaction with the SMN protein. RNA 7(11):1531–1542

Brasher SV et al. (2000) The structure of mouse HP1 suggests a unique mode of single peptide recognition by the shadow chromo domain dimer. Embo J 19(7):1587–1597

Callebaut I, Courvalin JC, Mornon JP (1999) The BAH (bromo-adjacent homology) domain: a link between DNA methylation, replication and transcriptional regulation. FEBS Lett 446(1):189–193

Cao R et al. (2002) Role of histone H3 lysine 27 methylation in Polycomb-group silencing. Science 298(5595):1039–1043

Capili AD et al. (2001) Solution structure of the PHD domain from the KAP-1 corepressor: structural determinants for PHD, RING and LIM zinc-binding domains. Embo J 20(1–2):165–177

Chavali GB et al. (2005) Crystal structure of the ENT domain of human EMSY. J Mol Biol 350(5):964–973

Col E et al. (2001) The histone acetyltransferase, hGCN5, interacts with and acetylates the HIV transactivator, Tat. J Biol Chem 276(30):28179–28184

Cote J, Richard S (2005) Tudor domains bind symmetrical dimethylated arginines. J Biol Chem 280(31):28476–28483

Dhalluin C et al. (1999) Structure and ligand of a histone acetyltransferase bromodomain. Nature 399(6735):491–496

Dorr A et al. (2002) Transcriptional synergy between Tat and PCAF is dependent on the binding of acetylated Tat to the PCAF bromodomain. Embo J 21(11):2715–2723

Dutta S et al. (2001) The crystal structure of nucleoplasmin-core: implications for histone binding and nucleosome assembly. Mol Cell 8(4):841–853

Eberharter A et al. (2001) Acf1, the largest subunit of CHRAC, regulates ISWI-induced nucleosome remodelling. Embo J 20(14):3781–3788

Eberharter A et al. (2004) ACF1 improves the effectiveness of nucleosome mobilization by ISWI through PHD-histone contacts. Embo J 23(20):4029–4039

Edwards TA et al. (2005) Solution Structure of the Vts1 SAM Domain in the Presence of RNA. J Mol Biol (in press)

Ekblad CM et al. (2005) Binding of EMSY to HP1beta: implications for recruitment of HP1beta and BS69. EMBO Rep 6(7):675–680

Ekman D et al. (2005) Multi-domain proteins in the three kingdoms of life: orphan domains and other unassigned regions. J Mol Biol 348(1):231–243

Fischle W, Wang Y, Jacobs SA, Kim Y, Allis CD, Khorasanizadeh S (2003) Molecular basis for the discrimination of repressive methyl-lysine marks in histone H3 by Polycomb and HP1 chromodomains. Genes Dev 17(15):1870–1881

Friesen WJ et al. (2001) SMN, the product of the spinal muscular atrophy gene, binds preferentially to dimethylarginine-containing protein targets. Mol Cell 7(5):1111–1117

Ge YZ et al. (2004) Chromatin targeting of de novo DNA methyltransferases by the PWWP domain. J Biol Chem 279(24):25447–25454

Gibson TJ et al. (1998) The APECED polyglandular autoimmune syndrome protein, AIRE-1, contains the SAND domain and is probably a transcription factor. Trends Biochem Sci 23(7):242–244

Goodwin GH, Nicolas RH (2001) The BAH domain, polybromo and the RSC chromatin remodelling complex. Gene 268(1–2):1–7

Gozani O et al. (2003) The PHD finger of the chromatin-associated protein ING2 functions as a nuclear phosphoinositide receptor. Cell 114(1):99–111

Green JB et al. (2003) RNA recognition via the SAM domain of Smaug. Mol Cell 11(6):1537–1548

Grune T et al. (2003) Crystal structure and functional analysis of a nucleosome recognition module of the remodeling factor ISWI. Mol Cell 12(2):449–460

Halbach T, Scheer N, Werr W (2000) Transcriptional activation by the PHD finger is inhibited through an adjacent leucine zipper that binds 14-3-3 proteins. Nucleic Acids Res 28(18):3542–3550

Haynes SR et al. (1992) The bromodomain: a conserved sequence found in human, Drosophila and yeast proteins. Nucleic Acids Res 20(10):2603

Hennig L, Bouveret R, Gruissem W (2005) MSI1-like proteins: an escort service for chromatin assembly and remodeling complexes. Trends Cell Biol 15(6):295–302

Hou Z et al. (2005) Structural basis of the Sir1-origin recognition complex interaction in transcriptional silencing. Proc Natl Acad Sci USA 102(24):8489–8494

Hsu HC, Stillman B, Xu RM (2005) Structural basis for origin recognition complex 1 protein-silence information regulator 1 protein interaction in epigenetic silencing. Proc Natl Acad Sci USA 102(24):8519–8524

Hudson BP et al. (2000) Solution structure and acetyl-lysine binding activity of the GCN5 bromodomain. J Mol Biol 304:355–370

Huyen Y et al. (2004) Methylated lysine 79 of histone H3 targets 53BP1 to DNA double-strand breaks. Nature 432(7015):406–411

Ito T et al. (1996) ATP-facilitated chromatin assembly with a nucleoplasmin-like protein from Drosophila melanogaster. J Biol Chem 271(40):25041–25048

Ito T et al. (1999) ACF consists of two subunits, Acf1 and ISWI, that function cooperatively in the ATP-dependent catalysis of chromatin assembly. Genes Dev 13(12):1529–1539

Iwahara J et al. (2002) The structure of the Dead ringer-DNA complex reveals how AT-rich interaction domains (ARIDs) recognize DNA. Embo J 21(5):1197–1209

Jacobs SA, Khorasanizadeh S (2002) Structure of HP1 chromodomain bound to a lysine 9-methylated histone H3 tail. Science 295(5562):2080–2083

Jacobson RH et al. (2000) Structure and function of a human TAFII250 double bromodomain module. Science 288:1422–1425

Jeanmougin F et al. (1997) The bromodomain revisited. Trends Biochem Sci 22(5):151–153

Johnson PE, Donaldson LW (2006) RNA recognition by the Vts1p SAM domain. Nat Struct Mol Biol 13(2):177–178.

Kalkhoven E et al. (2002) The PHD type zinc finger is an integral part of the CBP acetyltransferase domain. Mol Cell Biol 22(7):1961–1970

Kalkhoven E et al. (2003) Loss of CBP acetyltransferase activity by PHD finger mutations in Rubinstein-Taybi syndrome. Hum Mol Genet 12(4):441–450

Kanno T et al. (2004) Selective recognition of acetylated histones by bromodomain proteins visualized in living cells. Mol Cell 13(1):33–43

Kasten M et al. (2004) Tandem bromodomains in the chromatin remodeler RSC recognize acetylated histone H3 Lys14. Embo J 23(6):1348–1359

Kim CA et al. (2002) The SAM domain of polyhomeotic forms a helical polymer. Nat Struct Biol 9(6):453–457

Kim CA et al. (2005) Structural organization of a Sex-comb-on-midleg/polyhomeotic copolymer. J Biol Chem 280(30):27769–27775

Kim S et al. (2004) Structure and DNA-binding sites of the SWI1 AT-rich interaction domain (ARID) suggest determinants for sequence-specific DNA recognition. J Biol Chem 279(16):16670–16676

Kodandapani R et al. (1996) A new pattern for helix-turn-helix recognition revealed by the PU.1 ETS-domain-DNA complex. Nature 380(6573):456–460

Kortschak RD, Tucker PW, Saint R (2000) ARID proteins come in from the desert. Trends Biochem Sci 25(6):294–299

Kwan AH et al. (2003) Engineering a protein scaffold from a PHD finger. Structure (Camb) 11(7):803–813

Lachner M et al. (2001) Methylation of histone H3 lysine 9 creates a binding site for HP1 proteins. Nature 410(6824):116–120

Ladurner AG et al. (2003) Bromodomains mediate an acetyl-histone encoded antisilencing function at heterochromatin boundaries. Mol Cell 11(2):365–376

Lu Z et al. (2002) The PHD domain of MEKK1 acts as an E3 ubiquitin ligase and mediates ubiquitination and degradation of ERK1/2. Mol Cell 9(5):945–956

Martinez-Campa C et al. (2004) Precise nucleosome positioning and the TATA box dictate requirements for the histone H4 tail and the bromodomain factor Bdf1. Mol Cell 15(1):69–81

Matangkasombut O, Buratowski S (2003) Different sensitivities of bromodomain factors 1 and 2 to histone H4 acetylation. Mol Cell 11(2):353–363

Maurer-Stroh S et al. (2003) The Tudor domain Royal Family: Tudor, plant Agenet, Chromo, PWWP and MBT domains. Trends Biochem Sci 28(2):69–74

Min J, Zhang Y, Xu RM (2003) Structural basis for specific binding of Polycomb chromodomain to histone H3 methylated at Lys 27. Genes Dev 17(15):1823–1828

Mujtaba S et al. (2002) Structural basis of lysine-acetylated HIV-1 Tat recognition by PCAF bromodomain. Mol Cell 9(3):575–586

Mujtaba S et al. (2004) Structural mechanism of the bromodomain of the coactivator CBP in p53 transcriptional activation. Mol Cell 13(2):251–263

Nakayama J et al. (2001) Role of histone H3 lysine 9 methylation in epigenetic control of heterochromatin assembly. Science 292(5514):110–113

Namboodiri VM et al. (2003) The crystal structure of Drosophila NLP-core provides insight into pentamer formation and histone binding. Structure (Camb) 11(2):175–186

Nameki N et al. (2005) Solution structure of the PWWP domain of the hepatoma-derived growth factor family. Protein Sci 14(3):756–764

Nielsen PR et al. (2002) Structure of the HP1 chromodomain bound to histone H3 methylated at lysine 9. Nature 416(6876):103–107

Nielsen PR et al. (2005) Structure of the chromo barrel domain from the MOF acetyltransferase. J Biol Chem 280(37):32326–32331

Noguchi E et al. (2003) Swi1 prevents replication fork collapse and controls checkpoint kinase Cds1. Mol Cell Biol 23(21):7861–7874

Oberstrass FC et al. (2006) Shape-specific recognition in the structure of the Vts1p SAM domain with RNA. Nat Struct Mol Biol 13(2):160–167.

Oliver AW et al. (2005) Crystal structure of the proximal BAH domain of the polybromo protein. Biochem J 389(Pt 3):657–664

Ornaghi P et al. (1999) The bromodomain of Gcn5p interacts in vitro with specific residues in the N terminus of histone H4. J Mol Biol 287(1):1–7

Owen DJ et al. (2000) The structural basis for the recognition of acetylated histone H4 by the bromodomain of histone acetyltransferase gcn5p. Embo J 19(22):6141–6149

Pak DT et al. (1997) Association of the origin recognition complex with heterochromatin and HP1 in higher eukaryotes. Cell 91(3):311–323

Pascual J et al. (2000) Structure of the PHD zinc finger from human Williams-Beuren syndrome transcription factor. J Mol Biol 304(5):723–729

Patsialou A, Wilsker D, Moran E (2005) DNA-binding properties of ARID family proteins. Nucleic Acids Res 33(1):66–80

Polesskaya A et al. (2001) Interaction between acetylated MyoD and the bromodomain of CBP and/or p300. Mol Cell Biol 21(16):5312–5320

Poot RA et al. (2000) HuCHRAC, a human ISWI chromatin remodelling complex contains hACF1 and two novel histone-fold proteins. Embo J 19(13):3377–3387

Pray-Grant MG et al. (2005) Chd1 chromodomain links histone H3 methylation with SAGA- and SLIK-dependent acetylation. Nature 433(7024):434–438

Qian C et al. (2005) Structure and chromosomal DNA binding of the SWIRM domain. Nat Struct Mol Biol 12(12):1078–1085

Qiao F, Bowie JU (2005) The many faces of SAM. Sci STKE 2005(286):re7

Qiu C et al. (2002) The PWWP domain of mammalian DNA methyltransferase Dnmt3b defines a new family of DNA-binding folds. Nat Struct Biol 9(3):217–224

Ragvin A et al. (2004) Nucleosome binding by the bromodomain and PHD finger of the transcriptional cofactor p300. J Mol Biol 337(4):773–788

Ramakrishnan V et al. (1993) Crystal structure of globular domain of histone H5 and its implications for nucleosome binding. Nature 362(6417):219–223

Roth SY, Denu JM, Allis CD (2001) Histone acetyltransferases. Annu Rev Biochem 70:81–120

Sathyamurthy A et al. (2003) Crystal structure of the malignant brain tumor (MBT) repeats in Sex Comb on Midleg-like 2 (SCML2). J Biol Chem 278(47):46968–46973

Selenko P et al. (2001) SMN tudor domain structure and its interaction with the Sm proteins. Nat Struct Biol 8(1):27–31

Shamay M et al. (2002) Hepatitis B virus pX interacts with HBXAP, a PHD finger protein to coactivate transcription. J Biol Chem 277(12):9982–9988

Shi Y et al. (2004) Histone demethylation mediated by the nuclear amine oxidase homolog LSD1. Cell 119(7):941–953

Simic R et al. (2003) Chromatin remodeling protein Chd1 interacts with transcription elongation factors and localizes to transcribed genes. Embo J 22(8):1846–1856

Slater LM, Allen MD, Bycroft M (2003) Structural variation in PWWP domains. J Mol Biol 330(3):571–576

Smith TF et al. (1999) The WD repeat: a common architecture for diverse functions. Trends Biochem Sci 24(5):181–185

Stec I et al. (2000) The PWWP domain: a potential protein–protein interaction domain in nuclear proteins influencing differentiation? FEBS Lett 473(1):1–5

Strahl BD, Allis CD (2000) The language of covalent histone modifications. Nature 403(6765):41–45

Surdo PL et al. (2003) Crystal structure and nuclear magnetic resonance analyses of the SAND domain from glucocorticoid modulatory element binding protein-1 reveals deoxyribonucleic acid and zinc binding regions. Mol Endocrinol 17(7):1283–1295

Thiru A et al. (2004) Structural basis of HP1/PXVXL motif peptide interactions and HP1 localisation to heterochromatin. Embo J 23(3):489–499

Turner BM (2002) Cellular memory and the histone code. Cell 111(3):285–291

Vermaak D et al. (1999) Functional analysis of the SIN3-histone deacetylase RPD3-RbAp48-histone H4 connection in the Xenopus oocyte. Mol Cell Biol 19(9):5847–5860

Verreault A et al. (1998) Nucleosomal DNA regulates the core-histone-binding subunit of the human Hat1 acetyltransferase. Curr Biol 8(2):96–108

Wang WK et al. (2003) Malignant brain tumor repeats: a three-leaved propeller architecture with ligand/peptide binding pockets. Structure (Camb) 11(7):775–789

Wilsker D et al. (2004) The DNA-binding properties of the ARID-containing subunits of yeast and mammalian SWI/SNF complexes. Nucleic Acids Res 32(4):1345–1353

Wilsker D et al. (2005) Nomenclature of the ARID family of DNA-binding proteins. Genomics 86(2):242–251

Wood LD et al. (2003) Small ubiquitin-like modifier conjugation regulates nuclear export of TEL, a putative tumor suppressor. Proc Natl Acad Sci USA 100(6):3257–3262

Wysocka J et al. (2005) WDR5 associates with histone H3 methylated at K4 and is essential for H3 K4 methylation and vertebrate development. Cell 121(6):859–872

Yu J et al. (2003) A SANT motif in the SMRT corepressor interprets the histone code and promotes histone deacetylation. Embo J 22(13):3403–3410

Zeng L, Zhou MM (2002) Bromodomain: an acetyl-lysine binding domain. FEBS Lett 513(1):124–128

Zhang Z et al. (2002) Structure and function of the BAH-containing domain of Orc1p in epigenetic silencing. Embo J 21(17):4600–4611

Zheng N et al. (1999) Structural basis of DNA recognition by the heterodimeric cell cycle transcription factor E2F-DP. Genes Dev 13(6):666–674

Zhou Y, Grummt I (2005) The PHD finger/bromodomain of NoRC interacts with acetylated histone H4K16 and is sufficient for rDNA silencing. Curr Biol 15(15):1434–1438

Zhu L et al. (2001) Dynamics of the Mrf-2 DNA-binding domain free and in complex with DNA. Biochemistry 40(31):9142–9150

Results Probl Cell Differ (41)
L. Brehon: Chromatin Dynamics in Cellular Function
DOI 10.1007/016/Published online: 12 April 2006
© Springer-Verlag Berlin Heidelberg 2006

The Generation and Recognition of Histone Methylation

Michael S. Torok · Patrick A. Grant (✉)

Department of Biochemistry and Molecular Genetics,
University of Virginia School of Medicine, Charlottesville, VA 22908, USA
pag9n@virginia.edu

Abstract The posttranslational modification of histone proteins via methylation has important functions in gene activation, transcriptional silencing, establishment of chromatin states, and likely many aspects of DNA metabolism. The identification of numerous effector protein domains with the capability of binding methylated histones has significantly advanced our understanding of how such histone modifications may exert their biological effects. Here, we summarize aspects of the generation of arginine and lysine methylation marks on core histones, the characterization of the protein modules that interact with them, and how histone methylation cross-talks with other modifications.

1
Introduction

Inherent within the DNA binary code is information describing how our genes are to be activated and inactivated and how those instructions dictate the molecular diversity found throughout each level of our bodies, from our organelles to our system of tissues. This is significant because the DNA in our cells, despite the tissue type, is invariably the same with few exceptions. However, it is the collective difference in gene expression that initiates the many cellular changes in both normal development and the pathological states. Consequently, if we can better understand how our genetic code is regulated, then perhaps we can discover what is functioning incorrectly in various diseases.

The nuclei in each cell successfully handle a DNA-size paradox. The paradox is this: a eukaryotic cell of the human body contains billions of base pairs of DNA, and when that DNA is stretched from end to end it can be as long as 2 m in length. However, the nucleus in those cells is only a few microns in diameter. Eukaryotic cells package DNA into chromatin in order to solve this macromolecular-size paradox. In turn, the chromatin structure regulates the access of DNA-acting factors to the DNA template. Consequently, all enzymatic activities involved in DNA metabolic processes, such as transcription, gene silencing, repair, elimination, imprinting, dosage compensation, replication, recombination, apoptosis, mitosis, and chromosome maintenance and stability, must act at the level of chromatin. Also, this has profound implications for heredity because genetic inheritance may not only be based on the binary code of DNA, but also on the chromatin environment which houses that code. Collectively, the term that best describes this kind of DNA regulation, via chromatin structure, is referred to as epigenetics (Turner 1993, 2000; Strahl and Allis, 2000). Epigenetics is a phenomenon that acts above the level of DNA, but directly influences DNA processes.

Chromatin serves as a major signaling hub within a cell. In signal transduction networks, a wide variety of intracellular and extracellular input signals are integrated onto a central platform (Pawson and Nash 2000). Chromatin is one such central platform where the very chromatin components are dynamic and can be modified with a variety of chemical groups, remodeled, and/or exchanged with variants. Also, a multitude of proteins and protein complexes that contain distinct binding modules can interact with the different chromatin modifications, and together serve as signal conduits adding an additional capacity for genomic regulation. Histone modifications and the effector proteins that recognize them regulate multiple, and likely all, aspects of DNA metabolism. As such, the remainder of this chapter will flow from a discussion of chromatin structure to the posttranslational modification of the histone components of chromatin by methylation and the recognition of this modification by effector proteins.

2
The Nucleosome and Chromatin Structure

The fundamental unit of chromatin is the nucleosome core particle, which has a molecular mass of approximately 206 kDa. Each nucleosome is composed of approximately 147 base pairs of DNA wrapped in a left-handed superhelix of 1.65 turns around an octamer of core histones H2A, H2B, H3, and H4 (Luger et al. 1997). An array of nucleosomes generally contains an additional histone, H1, which is also known as the "linker histone". Core histones range in size from 10 to 14 kDa, are essential proteins, and are found in equal molar stoichiometry in all eukaryotes. Core histones are highly conserved and are rich in positively charged lysine and arginine residues. They contain three unique domains called the histone fold, the histone fold extensions, and the histone tails. Histones H2A/H2B and H3/H4 heterodimerize through hydrophobic contacts between the histone fold domains in each histone protein. Within the nucleosome particle, one H3–H4 tetramer and two H2A–H2B dimers make a single octamer.

The *N*-terminal histone tails range in length from 16 to 44 amino acids, adopt no defined secondary structure, and are largely unresolved in the nucleosome crystal structure (Luger et al. 1997). Instead, they are believed to protrude away from the chromatin polymer and through the superhelical turns of the DNA in the nucleosome core particle. Some histones also contain *C*-terminal tails, the largest of which is on histone H2A. Histone tails, in particular, are subject to a multitude of posttranslational modifications (Allfrey et al. 1964; Jenuwein and Allis 2001; Peterson and Laniel 2004), and consequently function as nuclear signaling platforms where specific patterns of modifications coordinate specific DNA-templated processes.

2.1
Chromatin Domains

A large number of histone modifications are now known to be associated with certain chromatin states. In general, chromatin structure is functionally divided into two major domains: euchromatin and heterochromatin (Felsenfeld and Groudine 2003; Henikoff 2000). Biochemical, genetic, and cytological studies suggest that sections of euchromatin sustain active DNA-templated processes and sections of heterochromatin repress, or silence, active DNA-templated processes. Heterochromatin is characterized as compact, relatively inaccessible, rich in repetitive DNA, gene poor, late replicating, and refractory to recombination machinery (Katan-Khaykovich and Struhl 2005). It is stably localized to specific genomic sites primarily at the centromere, pericentromere, and telomere (Karpen and Allshire 1997) and in yeast, heterochromatin is also found at the mating-type loci and the rDNA (Guarente 1999; Loo and Rine 1995; Lowell and Pillus 1998). Functionally, heterochromatin structures are the root for diverse epigenetic phenomena such as position effect variegation (PEV) in *Drosophila* and female X chromosome inactivation in mammals. In PEV, euchromatin placed next to heterochromatin is silenced in a variegated fashion and this silencing can affect adjacent genes (Gottschling et al. 1990; Schotta et al. 2003; Wallrath and Elgin 1995). Variegation occurs when heterochromatic silencing proteins "spread" to euchromatic regions, subsequently silencing genomic regions.

2.2
Histone Modifications

Histone proteins were long considered monotonous DNA packaging proteins. However, a recent flood of studies focusing on the remodeling, exchange, and posttranslational modification of histones and biological readouts has proven this original notion completely incorrect. In its place is the new understanding that histone-containing nucleosomes are the fundamental regulatory unit of the genome.

Histones are subject to a high degree of posttranslational modification such as sumoylation, acetylation, methylation, ubiquitination, ADP ribosylation, glycosylation, deimination, citrullination, phosphorylation, and other less characterized modifications (Jenuwein and Allis 2001; Wang et al. 2004a). These modifications have been reported on the *N*-terminal and *C*-terminal tails of histones and on the histone fold motifs near critical DNA–histone interactions (Cosgrove et al. 2004). To date, modifications that occur on serine, lysine, and arginine residues within histones are the best characterized. Similar to cytoplasmic signaling events, serine and threonine residues are phosphate acceptor sites. Also, lysine residues can be modified in multiple ways. For instance, they can be acetylated, mono-, di-, or trimethylated, sumoylated, or ubiquitinated. Further, arginine residues can be mono- or (symmetrically or asymmetrically) dimethylated (Bannister et al. 2002; Shiio and Eisenman 2003). Functional characterization of specific histone methylation events is discussed in more detail below.

The modification of histone proteins potentially disrupts protein–protein interactions by electrostatic, steric, or structural alterations, which can influence higher order chromatin structures. Specifically, histone tail modifications can alter the DNA–histone tail, histone tail–histone tail, and histone tail–non-histone tail interactions (Cary et al. 1982; Garcia-Ramirez et al. 1992, 1995; Jenuwein and Allis 2001). However, only a few histone modifications have been shown to cause significant structural changes in chromatin (Peterson and Laniel 2004). Thus, there is a growing interest in how combinations of posttranslational histone modifications dictate DNA function, without the necessity of generating gross alterations in chromatin structure. Characterization of these modifications has led to the formation of the "histone code" or "epigenetic code" hypothesis, which states that the pattern of histone modifications within specific chromatin contexts can participate in directing specific nuclear processes and downstream DNA-metabolic events (Strahl and Allis 2000; Turner 1993, 2000). The identification of effector proteins with recognition specificity for certain histone modifications lends support to this hypothesis.

3
Histone Methylation

Proteins can be posttranslationally methylated, which commonly occurs on the carboxyl groups of glutamate, leucine, and isoprenylated cysteine, and on the side-chain nitrogen atom of lysines, arginines, and histidine. Multiple lines of evidence suggest that the histone proteins are methylated on lysines and arginines (Murray 1964) and in some instances at significant levels. For example, approximately 35% of H3 is methylated at lysine 4 (Sun and Allis 2002), and approximately 90% of H3 is methylated at lysine 79 in the budding yeast *Saccharomyces cerevisiae* (van Leeuwen et al. 2002). The enzymes that transfer a methyl group from *S*-adenosyl methionine (AdoMet) to histones are aptly called histone methyltransferases (HMTs) and can add up to three methyl groups to a lysine residue. Increasing the number of methyl moieties on a lysine correlates with an increase in the basicity of the lysine side chain (Baxter and Byvoet 1975; Rice and Allis 2001). The degree of methylation on any single lysine residue, whether mono-, di-, or trimethylated, can also influence the biological outcome (Santos-Rosa et al. 2002; Wang et al. 2003).

3.1
Lysine Methylation

Specific histone methylation events are correlated with various chromatin functions, including gene activation or repression (Rice and Allis 2001; Sims et al. 2003; Fig. 1). For instance, histone H3 lysine 9 methylation (K9Me) is linked to transcriptional repression, and correlates with heterochromatic regions and the inactive X chromosome in mammals (Bannister et al. 2001; Boggs et al. 2002; Nakayama et al. 2001; Noma et al. 2001; Peters et al. 2002). The enzyme that catalyzes K9Me in *Drosophila* was identified as a suppressor of variegation 3-9, or Su(var) 3-9, and was already associated with genomic

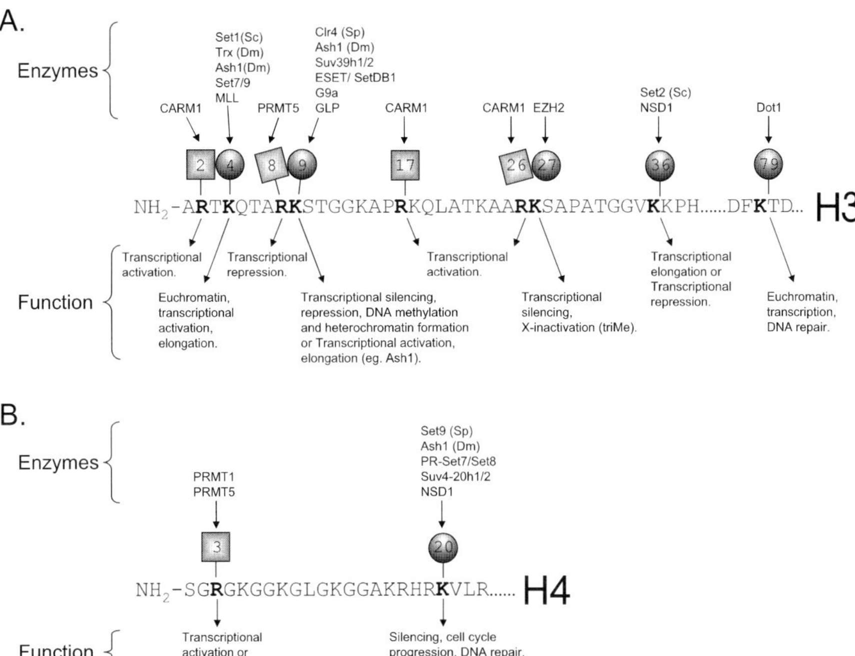

Fig. 1 Sites of histone methylation and their biological function. Sequences of the human H3 (**a**) or H4 (**b**) *N*-terminal tails, showing the amino acids subject to methylation in *bold*. Sites of arginine methylation are represented as *squares* and lysine methylation as *circles*. The enzymes implicated in generating methyl marks from mammals, the fruit fly *Drosophila melanogaster* (Dm), or the yeast species *S. cerevisiae* (Sc) and *S. pombe* (Sp) are listed above the sequences. Functions attributed to individual modifications are listed below each relevant amino acid

silencing. Su(var) 3-9 contains a SET (suppressor of variegation, enhancer of zeste, and trithorax) domain, which catalyzes histone methylation (Lachner and Jenuwein 2002; Rea et al. 2000). Interestingly, H3 K9Me cannot be detected in *S. cerevisiae*, but is detectable in *Schizosaccharomyces pombe* and higher eukaryotes (Strahl et al. 1999). This suggests that K9Me-mediated heterochromatin formation may be a regulatory mechanism that evolved in higher eukaryotes.

The trimethylation of histone H4 at lysine 20 (K20Me) is also associated with heterochromatin, mediated by the SET domain enzymes SUV4-20h1 and 2 (Rice et al. 2003; Schotta et al. 2004; Fig. 1B). However, mono- and dimethylated H4 K20Me and H3 K9Me are also associated with euchromatin, the significance of which is uncertain. It has been suggested that K20 monomethylation is required for cell-cycle progression, and accordingly levels are cell-cycle regulated. In this instance, the SET domain protein PR-Set7/SET8 has been implicated as the methyltransferase in this process (Karachentsev et al. 2005; Rice et al. 2002). Furthermore, the K20Me has been linked to DNA damage repair in *S. pombe* (see below). An emerging theme in the field of chromatin biology is that the degree of methylation of a particular histone residue can be associated with a profoundly different process.

The methylation of histone H3 at lysine 27 (K27Me) is also believed to be restricted to higher eukaryotes, and has also been linked to various forms of transcriptional silencing, including female X chromosome inactivation and genomic imprinting (Sims et al. 2003). K27Me is mediated by the SET domain protein enhancer of zeste (E(Z)) in fruit flies or its human homologue EZH2. Many additional SET proteins have been identified, such as Clr4, G9a, and ESET, which also methylate histone H3 lysine 9, while both yeast Set1 and human Set7/9 can methylate H3 lysine 4 (Briggs et al. 2001; Bryk et al. 2002; Lanzotti et al. 2002; Miller et al. 2001; Nagy et al. 2002; Nakayama et al. 2001; Nishioka et al. 2002; Roguev et al. 2003; Tachibana et al. 2001; Wang et al. 2001).

Histone H3 lysine 4 methylation (K4Me) is linked to active transcription, in part from studies using the transcriptionally active macronucleus of *Tetrahymena* and erythrocyte β-globin genes (Litt et al. 2001; Strahl et al. 1999). Chromatin immunoprecipitated (ChIP) DNA, when analyzed by microarray, revealed that H3 K4 trimethylation is found at the 5′ end of genes and highly correlates with high gene expression (Santos-Rosa et al. 2002; Sims et al. 2003; Bernstein et al. 2005). Further, ChIP experiments conducted with *S. pombe*, utilizing antisera against H3 K9Me or H3 K4Me, show that K9Me localizes predominantly to the heterochromatic mating loci, whereas K4Me localizes to the flanking euchromatin regions (Noma et al. 2001).

Interestingly, recent data show that Set1-mediated trimethylation of H3 lysine 4 is positively regulated by the RNA binding domain within the Set1 protein (Schlichter and Cairns 2005). This finding suggests a positive feedback loop for active transcription and RNA production. Still, evidence exists showing that Set1 may also mediate gene silencing in the case of rDNA because *set1* mutant yeast exhibits a loss of rDNA silencing (Bryk et al. 2002). However, this may represent an indirect consequence of set1 inactivation, since a loss of Set1 leads to decreased binding of Sir3 (silent information regulator 3) at heterochromatic sites (Santos-Rosa et al. 2004). While Sir3 is not directly required for rDNA silencing, it may suggest that Set1 is involved in maintaining a euchromatin environment or boundary elements that exclude the spreading of silencing factors from heterochromatin.

Other well-characterized lysine methylation events occur on H3 lysine 36 (K36Me) and H3 lysine 79 (K79Me) and are catalyzed by the Set2 and Dot1 enzymes, respectively (Lacoste et al. 2002; Ng et al. 2002; Strahl et al. 2002; van Leeuwen et al. 2002). K36 methylation by Set2 has been shown to have potentially opposing functions correlating with RNA polymerase II transcriptional elongation (Krogan et al. 2003; Li et al. 2003; Morris et al. 2005; Schaft et al. 2003) and transcriptional repression (Strahl et al. 2002). Dot1 is the only known histone lysine methyltransferase that does not contain a SET domain for catalysis. Overexpression or mutation of Dot1 disrupts telomeric silencing, suggesting that K79Me may also play a role in maintaining heterochromatic boundaries (van Leeuwen et al. 2002). In line with this reasoning, data show that hypomethylated H3 K79 localizes to silenced chromatin (Ng et al. 2003a). New evidence suggests another potential function for H3 K79Me where this mark has been shown to be important for survival after exposure to ionizing radiation in mammals (Game et al. 2005).

3.2
Arginine Methylation

Arginine residues within the histones can be methylated singly or doubly in one of two conformations, symmetric or asymmetric (Lee et al. 2005a). Histone arginine methylation is commonly linked to transcriptional activation and is important for embryonic development and cell differentiation. It is catalyzed by coactivator enzymes such as protein arginine methyltransferases (PRMTs) or coactivator-associated arginine methyltransferases (CARMs). PRMT4 and CARM1 methylate histone H3 at arginine residues 2, 17, and 26 (Chen et al. 1999; Schurter et al. 2001) and PRMT1 methylates arginine 3 of histone H4 (Schurter et al. 2001; Strahl et al. 2001). PRMT1- and CARM1-catalyzed asymmetric dimethylation is linked to gene activation; however, PRMT5-mediated symmetric dimethylation of arginine 3 of H4 and arginine 8 of H3 is associated with gene repression (Schurter et al. 2001; Pal et al. 2004). Thus, PRMT5 potentially antagonizes PRMT1 and CARM1 function, and particularly noteworthy is the fact that PRMT1 and PRMT5 both methylate the common residue at H4 arginine 3.

3.3
Histone Demethylation

Like reversible phosphorylation or acetylation of histones, methylated histones can be demethylated. The first lysine demethylase to be identified was the lysine-specific demethylase 1 (LSD1), with specificity for K4Me of histone H3. Characterization of LSD1 demonstrated that the enzyme does not cleave the methyl-ε-amino bond, but instead generates an unmodified lysine residue and formaldehyde via an amine oxidation reaction (Shi et al. 2004). This reaction requires the cofactor flavin adenine dinucleotide (FAD) and a protonated nitrogen and can demethylate mono- or dimethylated lysines, but not trimethylated lysines (Shi et al. 2004). This suggests that demethylation of trimethylated substrates is catalyzed by an unknown enzyme activity or is an enzymatically irreversible modification. The biological readout of demethylation reactions could potentially be to signal a loss of transcriptional activation, because LSD1 associates with histone deacetylases (HDACs) within the CoREST corepressor complex (Shi et al. 2004; Lee et al. 2005b). In fact, it was observed that downregulation of LSD1 results in increased K4Me and upregulation of LSD1 target genes. However, LSD1 has also been reported to function in hormone-dependent transcriptional activation (Metzger et al. 2005). In this report it was found that the association of LSD1 with androgen receptor confers H3-K9 demethylase activity upon the enzyme. Although the mechanism for mediating the substrate specificity of LSD1 awaits determination, the reports to date suggest that the same demethylase enzyme may be recruited in events of gene activation or repression, as determined by associating factors.

A novel family of H3-K36 demethylases has also recently been identified (Tsukada et al. 2005). In the presence of Fe(II) and α-ketogluterate, yeast and human JHDM1 (JmjC domain-containing histone demethylase 1) were found to preferentially demethylate a histone substrate dimethylated at K36. However, the enzyme was unable to demethylate an analogous trimethylated substrate. The JmjC domain-containing proteins are predicted to be metalloenzymes that regulate chromatin function. In this study, the JmjC domain of JHDM1 was found to be required for histone demethylase activity and was postulated to represent a signature motif of histone demethylases (Tsukada et al. 2005). Potentially, therefore, a large number of other uncharacterized JmjC proteins may represent histone demethylases and may include an elusive demethylase of trimethylated histones.

Histone arginine methylation is also reversible in a reaction known as deimination (Cuthbert et al. 2004; Wang et al. 2004b). Deimination is carried out by protein arginine demethylases (PADs). Unlike histone lysine demethylation, deimination is not truly reversible because the by-products of arginine demethylation are methylammonium and an altered amino acid citrulline. This suggests that any ultimate arginine residue "resetting" must be subsequently carried out through a mechanism involving histone replacement, such as would be achieved following DNA replication or through reconversion of citrulline to arginine.

4
Histone Modification Binding Proteins

Proteins fold into modular units that have a variety of structural and catalytic functions. Protein interaction motifs are generally smaller than 100 residues in length within the larger protein sequence (Bateman and Birney 2000). These can function by recruiting a substrate to a catalytic domain located elsewhere in the protein, interacting with other proteins, or modulating subcellular compartmentalization. Importantly, differences in the protein sequence alter recognition surfaces and allow for varying binding specificities. Further, proteins that contain multiple interaction modules may display increased binding specificities (Kuriyan and Cowburn 1997). A handful of protein domains have now been found to physically interact with specific histone methylation marks (Fig. 2), and thus are believed to transduce histone modifications in various functional pathways.

4.1
Chromodomains

One of the best-characterized histone interaction motifs is the chromatin organization modifier (chromo) domain. Chromodomains are found in various chromatin-acting proteins including the transcription-associated proteins Esa1 and Chd1, and the silencing-associated proteins Swi6, HP1, and Pc (Daniel et al. 2005). The first chromodomain structure to be solved was that of *Drosophila* HP1 (Jacobs and Khorasanizadeh 2002; Nielsen et al. 2002). Data derived from these studies revealed that aromatic amino acid side chains, within the HP1 chromodomain, associated with K9Me in the H3 tail. Specifically, π electrons in aromatic residues stabilized the methylammonium group

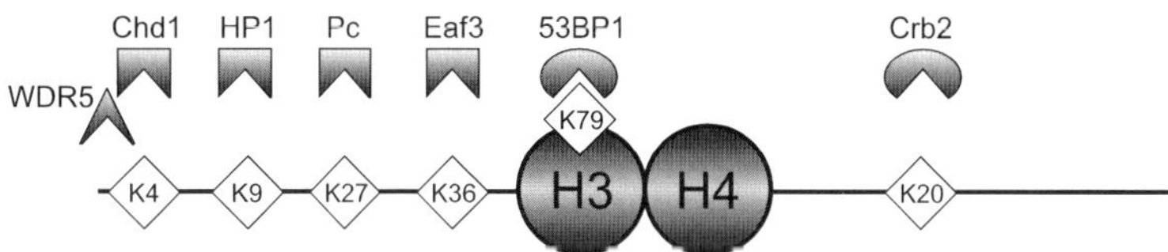

Fig. 2 Methyl lysine binding proteins. Illustration of the known effector proteins of methylated histones shown above their cognate methyl lysine binding sites on the histone H3 and H4 tails. Proteins are grouped according to the domain that they carry, namely WD40 (*triangular*), chromo (*circular*), or Tudor (*square*) domains

created when H3 K9 was methylated in what is referred to as an "aromatic cage" (Fischle et al. 2003; Jacobs and Khorasanizadeh 2002). These studies support earlier findings that H3 K9Me targets the silencing protein, HP1, to heterochromatin (Bannister et al. 2001; Lachner et al. 2001; Nakayama et al. 2001). Interestingly, HP1 may form homodimers and subsequently compact chromatin into higher order heterochromatin structures when bound to histone H3 K9Me (Thiru et al. 2004). Collectively this structure would contribute to an overall transcriptionally repressive chromatin environment.

Rather unexpectedly, another report has found that H3 K9 di- and trimethylation occur across the transcribed region of a number of active genes in mammalian cells, indicating that this methyl mark also functions in euchromatin (Vakoc et al. 2005). The HP1-γ isoform was found to dynamically associate with these transcriptionally active genes, and both methylation and HP1 association were dependent upon elongating RNA polymerase II. This suggests that K9Me and HP1 recruitment plays dual roles in both gene repression and transcription elongation.

The Polycomb (Pc) protein, a component of the Polycomb repressive complex-1 (PRC1), contains a chromodomain that has sequence similarity to HP1, and was therefore hypothesized to bind methylated lysines within histones. Multiple reports have demonstrated that Pc binds H3 K27Me (Cao et al. 2002; Czermin et al. 2002; Kuzmichev et al. 2002; Fischle et al. 2003), but these findings contrast with follow-up studies, which suggest that Pc can interact with both H3 K9Me and K27Me (Ringrose et al. 2004). Despite these disparities, the chromodomain-containing proteins HP1 and Pc function by binding methylated lysines, and establishing and maintaining silent chromatin (Fischle et al. 2003).

While HP1 and Pc are generally associated with heterochromatin, other chromodomain-containing proteins are associated with transcriptional activation. The yeast Chd1 (chromo-ATPase/helicase-DNA binding domain) protein has been shown to bind di- and trimethylated H3 K4Me peptides in vitro (Daniel et al. 2005; Pray-Grant et al. 2005). Lysine 4 is methylated by Set1, and K4Me highly correlates with active transcription (Ng et al. 2003b; Santos-Rosa et al. 2002; Sims et al. 2003; Bernstein et al. 2005). These biochemical studies are supported by genetic data showing that mutant Chd1 can suppress the growth defect of a *set1* mutant (Zhang et al. 2005). Interestingly, the Chd1 protein contains two chromodomain modules, suggesting that it may bind two methylated residues simultaneously. Chd1 has also been associated with transcriptional elongation and chromatin remodeling (Krogan et al. 2002; Simic et al. 2003; Tran et al. 2000), suggesting continuity between the active K4Me mark and the transcription-associated Chd1. Human CHD1 has been reported to specifically bind to methylated histone H3 substrates in a fashion that requires both functional chromodomains (Sims et al. 2005). It should be noted that in this report the authors failed to see yeast Chd1 binding to peptide substrates under the conditions used, so consequently further studies are needed to understand the differences between the yeast and human proteins.

Another yeast chromodomain protein, Eaf3, has been identified as a component of both the NuA4 HAT complex, involved in transcription activation and DNA damage repair, and a low molecular weight Rpd3 histone deacetylase complex, termed RpdS (Carozza et al. 2005; Koegh et al. 2005). Intriguingly, preferential deacetylation of coding regions, but not promoters, of yeast genes was found to require Eaf3 and Rpd3S and H3 K36Me mediated by Set2. Furthermore, the chromodomain of Eaf3 was found to be required for the recruitment of Rpd3S and deacetylation within the open reading frames studied, and Eaf3 chromodomain was found to interact with mono-, di-, and trimethylated H3 K36Me substrates. The function of Eaf3 as part of the Rpd3S complex is apparently to prevent internal aberrant transcripts within mRNA coding regions. Collectively these observations suggest that histone deactetylation is linked to patterns of K36Me, which in turn is dependent upon Set2 and RNA polymerase II *C*-terminal phosphorylation during transcriptional elongation. Rpd3 recognition and recruitment by K36Me would then enable the deacetylation and resetting of chromatin structure following elongation, which is important for subsequent correct start site transcription initiation. A problem that still needs to be resolved is how Eaf3 in the Rpd3S deacetylase complex is recruited to K36Me marks, while the potentially antagonistic NuA4 acetyltransferase complex is apparently not.

4.2
Tudor and Malignant Brain Tumor Domains

Methylated H3 lysine 79, like lysine 4, is localized to active chromatin regions (Ng et al. 2002; van Leeuwen et al. 2002). The K79Me modification is recognized by the DNA damage response protein 53BP1 (Huyen et al. 2004). This finding supports other studies showing that methylation of H3 K79 is important for survival after exposure to ionizing radiation in mammals (Game et al. 2005). Interestingly, 53BP1 does not contain a canonical chromodomain, but rather a tandem Tudor domain. This domain is required for 53BP1 to accumulate within nuclear foci at sites of double strand breaks. It has been postulated that DNA breaks cause histone structural changes that expose an otherwise masked K79Me mark, enabling 53BP1 binding. A similar mechanism has been suggested for recruiting Crb2, a homologue *S. pombe* cell-cycle checkpoint protein, to sites of histone H4 K20Me following DNA damage (Sanders et al. 2004).

Tudor domains are a member of a superfamily of protein folds including the structurally similar chromodomains, malignant brain tumor (MBT) domains, PWWP (conserved proline and tryptophan) domains, and Agenet domains (Maurer-Stroh et al. 2003). Tudor domains, like chromodomains, have been shown to interact with methylated proteins. In particular, Tudor domains bind methylated lysine and arginine residues with similar structures to that observed in K9Me-HP1 binding studies (Brahms et al. 2001; Friesen et al. 2001; Jacobs and Khorasanizadeh 2002; Nielsen et al. 2002; Selenko et al. 2001). It has also been determined that the MBT domain from the *Drosophila* l(3)mbt protein has affinity for mono- and dimethyl lysines (W. Fischle, S. Nimer, and C.D. Allis, personal communication), further expanding our knowledge of histone methyl binders.

4.3
WD40 Domain

A recent report demonstrates that the WDR5 protein can also associate with dimethylated H3 K4 nucleosomes (Wysocka et al. 2005). This is an interesting finding because WDR5 does not contain a chromodomain, but rather a WD40 propeller motif. The WD40 propeller comprises a seven-banded propeller with a cavity in the center. WDR5 is a component of the MLL1, MLL2, and hSET1 H3 K4 methyltransferase family of complexes, suggesting that WDR5 may be able to interact with the same chromatin modification that these SET enzymes catalyze (Wysocka et al. 2005). The imitation switch (ISWI) remodeling complex has also been shown to interact with H3 K4Me (Santos-Rosa et al. 2003), suggesting that a single methylated lysine can recruit multiple proteins. Nevertheless, additional histone modifications within the context of K4Me or the number of methyl groups on that lysine may influence specific recruitment of different chromatin-acting proteins.

5
Histone Modification Crosstalk with Methylation

Physiologically, each histone most likely contains a large assortment of modifications. Therefore, not surprisingly, histone modification crosstalk occurs and initial histone modifications play important roles in regulating other subsequent histone modification events.

Histone H2B ubiquitination can be carried out by the E2 enzyme Rad6/Ubc2 (Jentsch et al. 1987; Sharon et al. 1991; Sung et al. 1988). The broad functions of Rad6 remain unclear; however, analysis of Rad6 indicates that mutations in this enzyme yield ultraviolet radiation sensitivity and defects in meiosis, protein degradation, retrotransposition, and a loss of silencing at telomeres, rDNA, and *HM* loci (Bryk et al. 1997; Hochstrasser 1996; Huang et al. 1997). Structural studies suggest that Rad6 localization to transcriptionally active promoters requires the histone ubiquitin ligase Bre1 (Hwang et al. 2003; Wood et al. 2003).

In *S. cerevisiae*, the *C*-terminal tail of H2B is ubiquitinated on lysine 123 (Robzyk et al. 2000), and this modification is predicted to disrupt internucleosomal chromatin folding (White et al. 2001). Structurally, ubiquitination of H2B forms a large adduct onto the chromatin structure and therefore may greatly influence genomic dynamics. This modification is evolutionarily conserved and occurs in mammals on lysine 120 (Robzyk et al. 2000; West and Bonner 1980). In yeast approximately 5% of H2B is ubiquitinated (Robzyk et al. 2000; Sun and Allis 2002), and has been found to regulate transcription (Davie and Murphy 1990; Nickel et al. 1989; Zhang 2003), and to associate with elongating PolII (Xiao et al. 2005). Interestingly, ubiquitination of histone H2B at lysine 123 is required for H3 K4 and K79 methylation, but not for methylation of K36 (Briggs et al. 2002; Dover et al. 2002; Ng et al. 2002; Shahbazian et al. 2005; Sun and Allis 2002). Perhaps a distinct ubiquitination event is required for H3 K36Me; however, there is no evidence of such an activity.

Similar to the reversible nature of acetylation, methylation, and phosphorylation events, histone ubiquitination is also reversible. H2B lysine 123 is deubiquitinated by the SAGA and SAGA-like (SLIK/SALSA) Ubp8 enzyme (Daniel et al. 2004; Henry et al. 2003). SAGA and SLIK are two related and conserved histone acetyltransferase complexes involved in transcription of RNA polymerase II genes, particularly those involved in stress responses in yeast. They contain different groups of proteins, many of which are well characterized as having roles in transcription (Torok and Grant 2004). Study of the Ubp8 enzyme reveals that deubiquitination of H2B, like ubiquitination of H2B, influences H3 K4Me levels which in turn correlate with transcriptional activity. However, studies differ on exactly how Ubp8 affects K4Me, with one study showing that *ubp8*-deleted yeast exhibits a dramatic increase in mono-K4Me and a slight decrease in tri-K4Me from bulk histones (Daniel et al. 2004). At the *GAL1-10* gene upstream activating sequence (UAS), the transition of H3 K4 trimethylation that accompanies *GAL* gene expression is dependent on Ubp8. Another study suggests that *ubp8* mutant yeast has increases in tri-K4Me at the *GAL1-10* promoter and no affect on monomethylation (Henry et al. 2003). The differences in these studies may be explained by the use of different antisera or the locations where methylated status was monitored, i.e., UAS versus promoter. Nevertheless, Ubp8 regulates the methylation status of H3 lysine 4. Recent reports observe that H2B is also deubiquitinated by the Ubp10 enzyme (Emre et al. 2005; Gardner et al. 2005) and that *ubp10* yeast exhibits a slightly higher level of H3 K4 mono- and trimethylation (Gardner et al. 2005).

Chd1 was found to physically interact with the SLIK and SAGA histone acetyltransferase complexes, and to potentiate preferential acetylation of K4Me peptides by SLIK and acetylation of the *GAL1-10* UAS in vivo (Pray-Grant et al. 2005; Daniel et al. 2005). This links histone H3 K4Me and H3 acetylation and potentially explains the coordination of these methyl–acetyl marks (Bernstein et al. 2005). A study of the WDR5 H3 K4Me-binding protein found that the protein physically interacts with the MLL lysine 4 methyltransferase and the MOF1 histone H4 acetyltransferase. Both enzymes are required for optimal transcriptional activity of a target gene in vivo and in vitro (Dou et al. 2005), and WDR5 was required specifically for H3 K4 trimethylation (Wysocka et al. 2005). Again, these studies help to explain the coordination of H3 K4 trimethylation and H4 acetylation during gene activation, and it is quite plausible that H3-acetylated and H3 K4Me chromatin is most competent for transcription.

As discussed, studies establish that K9Me is linked to transcriptional repression (Wang et al. 2001). These findings may be explained by the aforementioned recruitment of HP1 to histone H3 methylated at K9, which is also associated with transcriptional repression (Jacobs and Khorasanizadeh 2002; Nielsen et al. 2002). Additionally, the methylation of DNA, important in gene silencing in a number of biological processes, has been found to either regulate or be regulated by K9Me (Sims et al. 2003; Rice and Allis 2001). Clearly, the coordination of both histone K9 and DNA methylation is an efficient means to silence chromatin and establish a stable epigenetic state. Other data suggest that K9Me actually functions in precluding posttranslational modifications that correlate with active chromatin environments. For instance, K9Me inhibits the p300-mediated acetylation of H3 lysines 14, 18, and 23 and H3 lysine 4 methylation (Wang et al. 2001). Furthermore, methylation of a particular lysine residue precludes acetylation of that same residue and vice versa, where acetylation of a particular lysine precludes the methylation of that same residue.

Interestingly, this type of histone crosstalk can occur in the opposite direction, where active chromatin inhibits modifications associated with repressive chromatin. H3 K4Me has been shown to inhibit K9Me by the Su(var) 3-9 methyltransferase and alternatively, to promote H3 acetylation by the p300 HAT (Wang et al. 2001). Also, the transcriptionally active modification, phosphorylated H3 Serine 10, inhibits K9Me in vitro (Li et al. 2002; Rea et al. 2000). Recent reports demonstrate that Serine 10 phosphorylation and K9Me are not mutually exclusive in vivo and that the phosphorylation of H3 Serine 10 can eject HP1 bound to the adjacent K9Me residue (Fischle et al. 2005; Hirota et al. 2005). Other data reveal that modifications associated with active chromatin influence the creation of additional active chromatin modifications. For instance, acetylation of H3 lysines 9 and 14 stimulates K4Me by the MLL methyltransferase complex (Milne et al. 2002) and H3 Serine 10 phosphorylation (Rea et al. 2000).

6
Conclusions and Future Perspectives

Taken together, collective chromatin modifications mediate biological phenomena and in principle give credence to the histone code hypothesis. The first component of the code involves the cellular signaling cascade to the chromatin level. The second component is the regulation of histone-modifying enzymes and subsequent modifications of histones, which occurs in specific patterns. The third component is that chromatin-acting effector proteins can subsequently bind specific histone modifications and transduce the nuclear response respective to the "signal in" cascade. Additionally, the code also hypothesizes that epigenetic information can be transmitted from one generation to the next via histone modification binders (Turner 2002). It is clear that much remains to be learned about the generation, regulation, and recognition of histone methylation, but the explosion of information in the past few years makes this an exciting scientific area. Already at this early stage it is known that some protein domains have the ability to bind to a particular modification mark. Also, from what investigators have learned from lysine 4 methylation, multiple protein domains can recognize a single modification. Moreover, within the chromodomain family there is a specific ability to preferentially recognize one methylated lysine over another.

It is highly anticipated that histone binders exist for methylated arginine residues and potentially have the ability to recognize symmetric versus asymmetric modifications. Furthermore, there are probably many more histone demethylases and binders of methyl lysine residues that await identification. Finally, it is likely that methyl binding proteins with a clear specificity for the degree of modification exist, i.e., mono- versus di- versus trimethylated lysine. For example, it has been postulated that an effector protein specifically recognizing H3 lysine trimethylation should act downstream of the WDR5 protein (Wysocka et al. 2005). What is already apparent from our current understanding of histone methylation is that the clue to their physiological significance is the recruitment and stabilization of regulatory proteins to specific chromatin locations during gene regulation or DNA damage repair. The next challenge is to understand how the various effector proteins transduce multiple modification patterns into the relevant biological pathway.

References

Allfrey VG, Faulkner R, Mirsky AE (1964) Acetylation and methylation of histones and their possible role in the regulation of RNA synthesis. Proc Natl Acad Sci USA 51:786–794

Bannister AJ, Zegerman P, Partridge JF, Miska EA, Thomas JO, Allshire RC, Kouzarides T (2001) Selective recognition of methylated lysine 9 on histone H3 by the HP1 chromo domain. Nature 410:120–124

Bannister AJ, Schneider R, Kouzarides T (2002) Histone methylation: dynamic or static? Cell 109:801–806

Bateman A, Birney F (2000) Searching databases to find protein domain organization. Adv Protein Chem 54:137–157

Baxter CS, Byvoet P (1975) CMR studies of protein modification. Progressive decrease in charge density at the epsilon-amino function of lysine with increasing methyl substitution. Biochem Biophys Res Commun 64:514–518

Bernstein BE, Kamal M, Lindblad-Toh K, Bekiranov S, Bailey DK, Huebert DJ, McMahon S, Karlsson EK, Kulbokas EJ, Gingeras TR, Schreiber SL, Lander ES (2005) Genomic maps and comparative analysis of histone modifications in human and mouse. Cell 120:169–181

Boggs BA, Cheung P, Heard E, Spector DL, Chinault AC, Allis CD (2002) Differentially methylated forms of histone H3 show unique association patterns with inactive human X chromosomes. Nat Genet 30:73–76

Brahms H, Meheus L, de Brabandere V, Fischer U, Luhrmann R (2001) Symmetrical dimethylation of arginine residues in spliceosomal Sm protein B/B' and the Sm-like protein LSm4, and their interaction with the SMN protein. RNA 7:1531–1542

Briggs SD, Bryk M, Strahl BD, Cheung WL, Davie JK, Dent SY, Winston F, Allis CD (2001) Histone H3 lysine 4 methylation is mediated by Set1 and required for cell growth and rDNA silencing in Saccharomyces cerevisiae. Genes Dev 15:3286–3295

Briggs SD, Xiao T, Sun ZW, Caldwell JA, Shabanowitz J, Hunt DF, Allis CD, Strahl BD (2002) Gene silencing: trans-histone regulatory pathway in chromatin. Nature 418:498

Bryk M, Banerjee M, Murphy M, Knudsen KE, Garfinkel DJ, Curcio MJ (1997) Transcriptional silencing of Ty1 elements in the RDN1 locus of yeast. Genes Dev 11:255–269

Bryk M, Briggs SD, Strahl BD, Curcio MJ, Allis CD, Winston F (2002) Evidence that Set1, a factor required for methylation of histone H3, regulates rDNA silencing in S. cerevisae by a Sir2-independent mechanism. Curr Biol 12:165–170

Cao R, Wang L, Wang H, Xia L, Erdjument-Bromage H, Tempst P, Jones RS, Zhang Y (2002) Role of histone H3 lysine 27 methylation in Polycomb-group silencing. Science 298:1039–1043

Carrozza MJ, Li B, Florens L, Suganuma T, Swanson SK, Lee KK, Shia WJ, Anderson S, Yates J, Washburn MP, Workman JL (2005) Histone H3 methylation by Set2 directs deacetylation of coding regions by Rpd3S to suppress spurious intragenic transcription. Cell 123:581–592

Cary PD, Crane-Robinson C, Bradbury EM, Dixon GH (1982) Effect of acetylation on the binding of N-terminal peptides of histone H4 to DNA. Eur J Biochem 127:137–143

Chen D, Ma H, Hong H, Koh SS, Huang SM, Schurter BT, Aswad DW, Stallcup MR (1999) Regulation of transcription by a protein methyltransferase. Science 284:2174–2177

Cosgrove MS, Boeke JD, Wolberger C (2004) Regulated nucleosome mobility and the histone code. Nat Struct Mol Biol 11:1037–1043

Cuthbert GL, Daujat S, Snowden AW, Erdjument-Bromage H, Hagiwara T, Yamada M, Schneider R, Gregory PD, Tempst P, Bannister AJ, Kouzarides T (2004) Histone deimination antagonizes arginine methylation. Cell 118:545–553

Czermin B, Melfi R, McCabe D, Seitz V, Imhof A, Pirrotta V (2002) Drosophila enhancer of Zeste/ESC complexes have a histone H3 methyltransferase activity that marks chromosomal Polycomb sites. Cell 111:185–196

Daniel JA, Torok MS, Sun ZW, Schieltz D, Allis CD, Yates JR 3rd, Grant PA (2004) Deubiquitination of histone H2B by a yeast acetyltransferase complex regulates transcription. J Biol Chem 279:1867–1871

Daniel JA, Pray-Grant MG, Grant PA (2005) Effector proteins for methylated histones: an expanding family. Cell Cycle 4:919–926

Davie JR, Murphy LC (1990) Level of ubiquitinated histone H2B in chromatin is coupled to ongoing transcription. Biochemistry 29:4752–4757

Dou Y, Milne TA, Tackett AJ, Smith ER, Fukuda A, Wysocka J, Allis CD, Chait BT, Hess JL, Roeder RG (2005) Physical association and coordinate function of the H3 K4 methyltransferase MLL1 and the H4 K16 acetyltransferase MOF. Cell 121:873–885

Dover J, Schneider J, Tawiah-Boateng MA, Wood A, Dean K, Johnston M, Shilatifard A (2002) Methylation of histone H3 by COMPASS requires ubiquitination of histone H2B by Rad6. J Biol Chem 277:28368–28371

Emre NC, Ingvarsdottir K, Wyce A, Wood A, Krogan NJ, Henry KW, Li K, Marmorstein R, Greenblatt JF, Shilatifard A, Berger SL (2005) Maintenance of low histone ubiquitylation by Ubp10 correlates with telomere-proximal Sir2 association and gene silencing. Mol Cell 17:585–594

Felsenfeld G, Groudine M (2003) Controlling the double helix. Nature 421:448–453

Fischle W, Wang Y, Jacobs SA, Kim Y, Allis CD, Khorasanizadeh S (2003) Molecular basis for the discrimination of repressive methyl-lysine marks in histone H3 by Polycomb and HP1 chromodomains. Genes Dev 17:1870–1881

Fischle W, Tseng BS, Dormann HL, Ueberheide BM, Garcia BA, Shabanowitz J, Hunt DF, Funabiki H, Allis CD (2005) Regulation of HP1-chromatin binding by histone H3 methylation and phosphorylation. Nature 438:1116–1122

Friesen WJ, Massenet S, Paushkin S, Wyce A, Dreyfuss G (2001) SMN, the product of the spinal muscular atrophy gene, binds preferentially to dimethylarginine-containing protein targets. Mol Cell 7:1111–1117

Game JC, Williamson MS, Baccari C (2005) X-ray survival characteristics and genetic analysis for nine Saccharomyces deletion mutants that show altered radiation sensitivity. Genetics 169:51–63

Garcia-Ramirez M, Dong F, Ausio J (1992) Role of the histone tails in the folding of oligonucleosomes depleted of histone H1. J Biol Chem 267:19587–19595

Garcia-Ramirez M, Rocchini C, Ausio J (1995) Modulation of chromatin folding by histone acetylation. J Biol Chem 270:17923–17928

Gardner RG, Nelson ZW, Gottschling DE (2005) Ubp10/Dot4p regulates the persistence of ubiquitinated histone H2B: distinct roles in telomeric silencing and general chromatin. Mol Cell Biol 25:6123–6139

Gottschling DE, Aparicio OM, Billington BL, Zakian VA (1990) Position effect at S. cerevisiae telomeres: reversible repression of Pol II transcription. Cell 63:751–762

Guarente L (1999) Diverse and dynamic functions of the Sir silencing complex. Nat Genet 23:281–285

Henikoff S (2000) Heterochromatin function in complex genomes. Biochim Biophys Acta 1470:1–8

Henry KW, Wyce A, Lo WS, Duggan LJ, Emre NC, Kao CF, Pillus L, Shilatifard A, Osley MA, Berger SL (2003) Transcriptional activation via sequential histone H2B ubiquitylation and deubiquitylation, mediated by SAGA-associated Ubp8. Genes Dev 17:2648–2663

Hirota T, Lipp JJ, Toh BH, Peters JM (2005) Histone H3 serine 10 phosphorylation by Aurora B causes HP1 dissociation from heterochromatin. Nature 438:1176–1180

Hochstrasser M (1996) Ubiquitin-dependent protein degradation. Annu Rev Genet 30:405–439

Huang L, Zhang W, Roth SY (1997) Amino termini of histones H3 and H4 are required for a1-alpha2 repression in yeast. Mol Cell Biol 17:6555–6562

Huyen Y, Zgheib O, Ditullio RA Jr, Gorgoulis VG, Zacharatos P, Petty TJ, Sheston EA, Mellert HS, Stavridi ES, Halazonetis TD (2004) Methylated lysine 79 of histone H3 targets 53BP1 to DNA double-strand breaks. Nature 432:406–411

Hwang WW, Venkatasubrahmanyam S, Ianculescu AG, Tong A, Boone C, Madhani HD (2003) A conserved RING finger protein required for histone H2B monoubiquitination and cell size control. Mol Cell 11:261–266

Jacobs SA, Khorasanizadeh S (2002) Structure of HP1 chromodomain bound to a lysine 9-methylated histone H3 tail. Science 295:2080–2083

Jentsch S, McGrath JP, Varshavsky A (1987) The yeast DNA repair gene RAD6 encodes a ubiquitin-conjugating enzyme. Nature 329:131–134

Jenuwein T, Allis CD (2001) Translating the histone code. Science 293:1074–1080

Karachentsev D, Sarma K, Reinberg D, Steward R (2005) PR-Set7-dependent methylation of histone H4 Lys 20 functions in repression of gene expression and is essential for mitosis. Genes Dev 19:431–435

Karpen GH, Allshire RC (1997) The case for epigenetic effects on centromere identity and function. Trends Genet 13:489–496

Katan-Khaykovich Y, Struhl K (2005) Heterochromatin formation involves changes in histone modifications over multiple cell generations. Embo J 24:2138–2149

Keogh MC, Kurdistani SK, Morris SA, Ahn SH, Podolny V, Collins SR, Schuldiner M, Chin K, Punna T, Thompson NJ, Boone C, Emili A, Weissman JS, Hughes TR, Strahl BD, Grunstein M, Greenblatt JF, Buratowski S, Krogan NJ (2005) Cotranscriptional set2 methylation of histone H3 lysine 36 recruits a repressive Rpd3 complex. Cell 123:593–605

Krogan NJ, Kim M, Ahn SH, Zhong G, Kobor MS, Cagney G, Emili A, Shilatifard A, Buratowski S, Greenblatt JF (2002) RNA polymerase II elongation factors of Saccharomyces cerevisiae: a targeted proteomics approach. Mol Cell Biol 22:6979–6992

Krogan NJ, Kim M, Tong A, Golshani A, Cagney G, Canadien V, Richards DP, Beattie BK, Emili A, Boone C et al (2003) Methylation of histone H3 by Set2 in Saccharomyces cerevisiae is linked to transcriptional elongation by RNA polymerase II. Mol Cell Biol 23:4207–4218

Kuriyan J, Cowburn D (1997) Modular peptide recognition domains in eukaryotic signaling. Annu Rev Biophys Biomol Struct 26:259–288

Kuzmichev A, Nishioka K, Erdjument-Bromage H, Tempst P, Reinberg D (2002) Histone methyltransferase activity associated with a human multiprotein complex containing the Enhancer of Zeste protein. Genes Dev 16:2893–2905

Lachner M, Jenuwein T (2002) The many faces of histone lysine methylation. Curr Opin Cell Biol 14:286–298

Lachner M, O'Carroll D, Rea S, Mechtler K, Jenuwein T (2001) Methylation of histone H3 lysine 9 creates a binding site for HP1 proteins. Nature 410:116–120

Lacoste N, Utley RT, Hunter JM, Poirier GG, Cote J (2002) Disruptor of telomeric silencing-1 is a chromatin-specific histone H3 methyltransferase. J Biol Chem 277:30421–30424

Lanzotti DJ, Kaygun H, Yang X, Duronio RJ, Marzluff WF (2002) Developmental control of histone mRNA and dSLBP synthesis during Drosophila embryogenesis and the role of dSLBP in histone mRNA 3′ end processing in vivo. Mol Cell Biol 22:2267–2282

Lee DY, Teyssier C, Strahl BD, Stallcup MR (2005a) Role of protein methylation in regulation of transcription. Endocr Rev 26:147–170

Lee MG, Wynder C, Cooch N, Shiekhattar R (2005b) An essential role for CoREST in nucleosomal histone 3 lysine 4 demethylation. Nature 437:432–435

Li B, Howe L, Anderson S, Yates JR III, Workman JL (2003) The Set2 histone methyltransferase functions through the phosphorylated carboxyl-terminal domain of RNA polymerase II J Biol Chem 278:8897–8903

Li J, Lin Q, Yoon HG, Huang ZQ, Strahl BD, Allis CD, Wong J (2002) Involvement of histone methylation and phosphorylation in regulation of transcription by thyroid hormone receptor. Mol Cell Biol 22:5688–5697

Litt MD, Simpson M, Gaszner M, Allis CD, Felsenfeld G (2001) Correlation between histone lysine methylation and developmental changes at the chicken beta-globin locus. Science 293:2453–2455

Loo S, Rine J (1995) Silencing and heritable domains of gene expression. Annu Rev Cell Dev Biol 11:519–548

Lowell JE, Pillus L (1998) Telomere tales: chromatin, telomerase and telomere function in Saccharomyces cerevisiae. Cell Mol Life Sci 54:32–49

Luger K, Mader AW, Richmond RK, Sargent DF, Richmond TJ (1997) Crystal structure of the nucleosome core particle at 2.8 Å resolution. Nature 389:251–260

Maurer-Stroh S, Dickens NJ, Hughes-Davies L, Kouzarides T, Eisenhaber F, Ponting CP (2003) The Tudor domain royal family: Tudor, plant Agenet, Chromo, PWWP and MBT domains. Trends Biochem Sci 28:69–74

Metzger E, Wissmann M, Yin N, Muller JM, Schneider R, Peters AH, Gunther T, Buettner R, Schule R (2005) LSD1 demethylates repressive histone marks to promote androgen-receptor-dependent transcription. Nature 437:436–439

Miller T, Krogan NJ, Dover J, Erdjument-Bromage H, Tempst P, Johnston M, Greenblatt JF, Shilatifard A (2001) COMPASS: a complex of proteins associated with a trithorax-related SET domain protein. Proc Natl Acad Sci USA 98:12902–12907

Milne TA, Briggs SD, Brock HW, Martin ME, Gibbs D, Allis CD, Hess JL (2002) MLL targets SET domain methyltransferase activity to Hox gene promoters. Mol Cell 10:1107–1117

Morris SA, Shibata Y, Noma K, Tsukamoto Y, Warren E, Temple B, Grewal SI, Strahl BD (2005) Histone H3 K36 methylation is associated with transcription elongation in Schizosaccharomyces pombe. Eukaryot Cell 4:1446–1454

Murray K (1964) The occurrence of ε-N-methyl lysine in histones. Biochemistry 3:10–15

Nagy PL, Griesenbeck J, Kornberg RD, Cleary ML (2002) A trithorax-group complex purified from Saccharomyces cerevisiae is required for methylation of histone H3. Proc Natl Acad Sci USA 99:90–94

Nakayama J, Rice JC, Strahl BD, Allis CD, Grewal SI (2001) Role of histone H3 lysine 9 methylation in epigenetic control of heterochromatin assembly. Science 292:110–113

Ng HH, Xu RM, Zhang Y, Struhl K (2002) Ubiquitination of histone H2B by Rad6 is required for efficient Dot1-mediated methylation of histone H3 lysine 79. J Biol Chem 277:34655–34657

Ng HH, Ciccone DN, Morshead KB, Oettinger MA, Struhl K (2003a). Lysine-79 of histone H3 is hypomethylated at silenced loci in yeast and mammalian cells: a potential mechanism for position-effect variegation. Proc Natl Acad Sci USA 100:1820–1825

Ng HH, Robert F, Young RA, Struhl K (2003b). Targeted recruitment of Set1 histone methylase by elongating Pol II provides a localized mark and memory of recent transcriptional activity. Mol Cell 11:709–719

Nickel BE, Allis CD, Davie JR (1989) Ubiquitinated histone H2B is preferentially located in transcriptionally active chromatin. Biochemistry 28:958–963

Nielsen PR, Nietlispach D, Mott HR, Callaghan J, Bannister A, Kouzarides T, Murzin AG, Murzina NV, Laue ED (2002) Structure of the HP1 chromodomain bound to histone H3 methylated at lysine 9. Nature 416:103–107

Nishioka K, Rice JC, Sarma K, Erdjument-Bromage H, Werner J, Wang Y, Chuikov S, Valenzuela P, Tempst P, Steward R et al (2002) PR-Set7 is a nucleosome-specific methyltransferase that modifies lysine 20 of histone H4 and is associated with silent chromatin. Mol Cell 9:1201–1213

Noma K, Allis CD, Grewal SI (2001) Transitions in distinct histone H3 methylation patterns at the heterochromatin domain boundaries. Science 293:1150–1155

Pal S, Vishwanath SN, Erdjument-Bromage H, Tempst P, Sif S (2004) Human SWI/SNF-associated PRMT5 methylates histone H3 arginine 8 and negatively regulates expression of ST7 and NM23 tumor suppressor genes. Mol Cell Biol 24:9630–9645

Pawson T, Nash P (2000) Protein–protein interactions define specificity in signal transduction. Genes Dev 14:1027–1047

Peters AH, Mermoud JE, O'Carroll D, Pagani M, Schweizer D, Brockdorff N, Jenuwein T (2002) Histone H3 lysine 9 methylation is an epigenetic imprint of facultative heterochromatin. Nat Genet 30:77–80

Peterson CL, Laniel MA (2004) Histones and histone modifications. Curr Biol 14:R546–R551

Pray-Grant MG, Daniel JA, Schieltz D, Yates JR III, Grant PA (2005) Chd1 chromodomain links histone H3 methylation with SAGA- and SLIK-dependent acetylation. Nature 433:434–438

Rea S, Eisenhaber F, O'Carroll D, Strahl BD, Sun ZW, Schmid M, Opravil S, Mechtler K, Ponting CP, Allis CD, Jenuwein T (2000) Regulation of chromatin structure by site-specific histone H3 methyltransferases. Nature 406:593–599

Rice JC, Allis CD (2001) Histone methylation versus histone acetylation: new insights into epigenetic regulation. Curr Opin Cell Biol 13:263–273

Rice JC, Nishioka K, Sarma K, Steward R, Reinberg D, Allis CD (2002) Mitotic-specific methylation of histone H4 Lys 20 follows increased PR-Set7 expression and its localization to mitotic chromosomes. Genes Dev 16:2225–2230

Rice JC, Briggs SD, Ueberheide B, Barber CM, Shabanowitz J, Hunt DF, Shinkai Y, Allis CD (2003) Histone methyltransferases direct different degrees of methylation to define distinct chromatin domains. Mol Cell 12:1591–1598

Ringrose L, Ehret H, Paro R (2004) Distinct contributions of histone H3 lysine 9 and 27 methylation to locus-specific stability of polycomb complexes. Mol Cell 16:641–653

Robzyk K, Recht J, Osley MA (2000) Rad6-dependent ubiquitination of histone H2B in yeast. Science 287:501–504

Roguev A, Schaft D, Shevchenko A, Aasland R, Stewart AF (2003) High conservation of the Set1/Rad6 axis of histone 3 lysine 4 methylation in budding and fission yeasts. J Biol Chem 278:8487–8493

Sanders SL, Portoso M, Mata J, Bahler J, Allshire RC, Kouzarides T (2004) Methylation of histone H4 lysine 20 controls recruitment of Crb2 to sites of DNA damage. Cell 119:603–614

Santos-Rosa H, Schneider R, Bannister AJ, Sherriff J, Bernstein BE, Emre NC, Schreiber SL, Mellor J, Kouzarides T (2002) Active genes are tri-methylated at K4 of histone H3. Nature 419:407–411

Santos-Rosa H, Schneider R, Bernstein BE, Karabetsou N, Morillon A, Weise C, Schreiber SL, Mellor J, Kouzarides T (2003) Methylation of histone H3 K4 mediates association of the Isw1p ATPase with chromatin. Mol Cell 12:1325–1332

Santos-Rosa H, Bannister AJ, Dehe PM, Geli V, Kouzarides T (2004) Methylation of H3 lysine 4 at euchromatin promotes Sir3p association with heterochromatin. J Biol Chem 279:47506–47512

Schaft D, Roguev A, Kotovic KM, Shevchenko A, Sarov M, Shevchenko A, Neugebauer KM, Stewart AF (2003) The histone 3 lysine 36 methyltransferase, SET2, is involved in transcriptional elongation. Nucleic Acids Res 31:2475–2482

Schlichter A, Cairns BR (2005) Histone trimethylation by Set1 is coordinated by the RRM, autoinhibitory, and catalytic domains. Embo J 24:1222–1231

Schotta G, Ebert A, Dorn R, Reuter G (2003) Position-effect variegation and the genetic dissection of chromatin regulation in Drosophila. Semin Cell Dev Biol 14:67–75

Schotta G, Lachner M, Sarma K, Ebert A, Sengupta R, Reuter G, Reinberg D, Jenuwein T (2004) A silencing pathway to induce H3-K9 and H4-K20 trimethylation at constitutive heterochromatin. Genes Dev 18:1251–1262

Schurter BT, Koh SS, Chen D, Bunick GJ, Harp JM, Hanson BL, Henschen-Edman A, Mackay DR, Stallcup MR, Aswad DW (2001) Methylation of histone H3 by coactivator-associated arginine methyltransferase 1. Biochemistry 40:5747–5756

Selenko P, Sprangers R, Stier G, Buhler D, Fischer U, Sattler M (2001) SMN tudor domain structure and its interaction with the Sm proteins. Nat Struct Biol 8:27–31

Shahbazian MD, Zhang K, Grunstein M (2005) Histone H2B ubiquitylation controls processive methylation but not monomethylation by Dot1 and Set1. Mol Cell 19:271–277

Sharon G, Raboy B, Parag HA, Dimitrovsky D, Kulka RG (1991) RAD6 gene product of Saccharomyces cerevisiae requires a putative ubiquitin protein ligase (E3) for the ubiquitination of certain proteins. J Biol Chem 266:15890–15894

Shi Y, Lan F, Matson C, Mulligan P, Whetstine JR, Cole PA, Casero RA (2004) Histone demethylation mediated by the nuclear amine oxidase homolog LSD1. Cell 119:941–953

Shiio Y, Eisenman RN (2003) Histone sumoylation is associated with transcriptional repression. Proc Natl Acad Sci USA 100:13225–13230

Simic R, Lindstrom DL, Tran HG, Roinick KL, Costa PJ, Johnson AD, Hartzog GA, Arndt KM (2003) Chromatin remodeling protein Chd1 interacts with transcription elongation factors and localizes to transcribed genes. Embo J 22:1846–1856

Sims RJ, Nishioka K, Reinberg D (2003) Histone lysine methylation: a signature for chromatin function. Trends Genet 19:629–639

Sims RJ, Chen CF, Santos-Rosa H, Kouzarides T, Patel SS, Reinberg D (2005) Human but not yeast CHD1 binds directly and selectively to histone H3 methylated at lysine 4 via its tandem chromodomains. J Biol Chem 280:41789–41792

Strahl BD, Allis CD (2000) The language of covalent histone modifications. Nature 403:41–45

Strahl BD, Ohba R, Cook RG, Allis CD (1999) Methylation of histone H3 at lysine 4 is highly conserved and correlates with transcriptionally active nuclei in Tetrahymena. Proc Natl Acad Sci USA 96:14967–14972

Strahl BD, Briggs SD, Brame CJ, Caldwell JA, Koh SS, Ma H, Cook RG, Shabanowitz J, Hunt DF, Stallcup MR, Allis CD (2001) Methylation of histone H4 at arginine 3 occurs in vivo and is mediated by the nuclear receptor coactivator PRMT1. Curr Biol 11:996–1000

Strahl BD, Grant PA, Briggs SD, Sun ZW, Bone JR, Caldwell JA, Mollah S, Cook RG, Shabanowitz J, Hunt DF, Allis CD (2002) Set2 is a nucleosomal histone H3-selective methyltransferase that mediates transcriptional repression. Mol Cell Biol 22:1298–1306

Sun ZW, Allis CD (2002) Ubiquitination of histone H2B regulates H3 methylation and gene silencing in yeast. Nature 418:104–108

Sung P, Prakash S, Prakash L (1988) The RAD6 protein of Saccharomyces cerevisiae polyubiquitinates histones, and its acidic domain mediates this activity. Genes Dev 2:1476–1485

Tachibana M, Sugimoto K, Fukushima T, Shinkai Y (2001) Set domain-containing protein, G9a, is a novel lysine-preferring mammalian histone methyltransferase with hyperactivity and specific selectivity to lysines 9 and 27 of histone H3. J Biol Chem 276:25309–25317

Thiru A, Nietlispach D, Mott HR, Okuwaki M, Lyon D, Nielsen PR, Hirshberg M, Verreault A, Murzina NV, Laue ED (2004) Structural basis of HP1/PXVXL motif peptide interactions and HP1 localisation to heterochromatin. Embo J 23:489–499

Torok MS, Grant PA (2004) Histone acetyltransferase proteins contribute to transcriptional processes at multiple levels. Adv Protein Chem 67:181–199

Tran HG, Steger DJ, Iyer VR, Johnson AD (2000) The chromo domain protein chd1p from budding yeast is an ATP-dependent chromatin-modifying factor. Embo J 19:2323–2331

Tsukada YI, Fang J, Erdjument-Bromage H, Warren ME, Borchers CH, Tempst P, Zhang Y (2006) Histone demethylation by a family of JmjC domain-containing proteins. Nature 439:811–816

Turner BM (1993) Decoding the nucleosome. Cell 75:5–8

Turner BM (2000) Histone acetylation and an epigenetic code. Bioessays 22:836–845

Turner BM (2002) Cellular memory and the histone code. Cell 111:285–291

Vakoc CR, Mandat SA, Olenchock BA, Blobel GA (2005) Histone H3 lysine 9 methylation and HP1gamma are associated with transcription elongation through mammalian chromatin. Mol Cell 19:381–391

van Leeuwen F, Gafken PR, Gottschling DE (2002) Dot1p modulates silencing in yeast by methylation of the nucleosome core. Cell 109:745–756

Wallrath LL, Elgin SC (1995) Position effect variegation in Drosophila is associated with an altered chromatin structure. Genes Dev 9:1263–1277

Wang H, Cao R, Xia L, Erdjument-Bromage H, Borchers C, Tempst P, Zhang Y (2001) Purification and functional characterization of a histone H3-lysine 4-specific methyltransferase. Mol Cell 8:1207–1217

Wang H, An W, Cao R, Xia L, Erdjument-Bromage H, Chatton B, Tempst P, Roeder RG, Zhang Y (2003) mAM facilitates conversion by ESET of dimethyl to trimethyl lysine 9 of histone H3 to cause transcriptional repression. Mol Cell 12:475–487

Wang Y, Fischle W, Cheung W, Jacobs S, Khorasanizadeh S, Allis CD (2004a) Beyond the double helix: writing and reading the histone code. Novartis Found Symp 259:3–17; discussion 17–21, pp 163–169

Wang Y, Wysocka J, Sayegh J, Lee YH, Perlin JR, Leonelli L, Sonbuchner LS, McDonald CH, Cook RG, Dou Y, Roeder RG, Clarke S, Stallcup MR, Allis CD, Coonrod SA (2004b). Human PAD4 regulates histone arginine methylation levels via demethylimination. Science 306:279–283

West MH, Bonner WM (1980) Histone 2B can be modified by the attachment of ubiquitin. Nucleic Acids Res 8:4671–4680

White CL, Suto RK, Luger K (2001) Structure of the yeast nucleosome core particle reveals fundamental changes in internucleosome interactions. Embo J 20:5207–5218

Wood A, Krogan NJ, Dover J, Schneider J, Heidt J, Boateng MA, Dean K, Golshani A, Zhang Y, Greenblatt JF et al (2003) Bre1, an E3 ubiquitin ligase required for recruitment and substrate selection of Rad6 at a promoter. Mol Cell 11:267–274

Wysocka J, Swigut T, Milne TA, Dou Y, Zhang X, Burlingame AL, Roeder RG, Brivanlou AH, Allis CD (2005) WDR5 associates with histone H3 methylated at K4 and is essential for H3 K4 methylation and vertebrate development. Cell 121:859–872

Xiao T, Kao CF, Krogan NJ, Sun ZW, Greenblatt JF, Osley MA, Strahl BD (2005) Histone H2B ubiquitylation is associated with elongating RNA polymerase II. Mol Cell Biol 25:637–651

Zhang L, Schroeder S, Fong N, Bentley DL (2005) Altered nucleosome occupancy and histone H3K4 methylation in response to transcriptional stress. Embo J 24:2379–2390

Zhang Y (2003) Transcriptional regulation by histone ubiquitination and deubiquitination. Genes Dev 17:2733–2740

Results Probl Cell Differ (41)
L. Brehon: Chromatin Dynamics in Cellular Function
DOI 10.1007/006/Published online: 10 February 2006
© Springer-Verlag Berlin Heidelberg 2006

Histone Ubiquitylation and the Regulation of Transcription

Mary Ann Osley (✉) · Alastair B. Fleming · Cheng-Fu Kao

Molecular Genetics and Microbiology, University of New Mexico School of Medicine,
Albuquerque, NM USA
mosley@salud.unm.edu, afleming@salud.unm.edu, ckao@salud.unm.edu

Abstract The small (76 amino acids) and highly conserved ubiquitin protein plays key roles in the physiology of eukaryotic cells. Protein ubiquitylation has emerged as one of the most important intracellular signaling mechanisms, and in 2004 the Nobel Prize was awarded to Aaron Ciechanower, Avram Hersko, and Irwin Rose for their pioneering studies of the enzymology of ubiquitin attachment. One of the most common features of protein ubiquitylation is the attachment of polyubiquitin chains (four or more ubiquitin moieties attached to each other), which is a widely used mechanism to target proteins for degradation via the 26S proteosome. However, it is noteworthy that the first ubiquitylated protein to be identified was histone H2A, to which a single ubiquitin moiety is most commonly attached. Following this discovery, other histones (H2B, H3, H1, H2A.Z, macroH2A), as well as many nonhistone proteins, have been found to be monoubiquitylated. The role of monoubiquitylation is still elusive because a single ubiquitin moiety is not sufficient to target proteins for turnover, and has been hypothesized to control the assembly or disassembly of multiprotein complexes by providing a protein-binding site. Indeed, a number of ubiquitin-binding domains have now been identified in both polyubiquitylated and monoubiquitylated proteins. Despite the early discovery of ubiquitylated histones, it has only been in the last five or so years that we have begun to understand how histone ubiquitylation is regulated and what roles it plays in the cell. This review will discuss current research on the factors that regulate the attachment and removal of ubiquitin from histones, describe the relationship of histone ubiquitylation to histone methylation, and focus on the roles of ubiquitylated histones in gene expression.

1
Regulation of Histone Ubiquitylation

Ubiquitylated histones have been estimated to account for 1–20% of total cellular histones – levels that are in part accounted for by the dynamic nature of histone ubiquitylation. The ubiquitin mark turns over continually throughout mitotic cell growth, and during mitosis the core histones are globally deubiquitylated at metaphase and reubiquitylated as cells enter anaphase (Goldknopf, Sudhakar et al. 1980; Seale 1981; Wu, Kohn et al. 1981; Mueller, Yasuda et al. 1985). Although ub-H2B is stable during mitosis in budding yeast, H2B is reversibly ubiquitylated during the transcription cycle in this organism (Henry, Wyce et al. 2003; Kao, Hillyer et al. 2004; Xiao, Kao et al. 2005). The dynamic regulation of histone ubiquitylation depends on numerous factors, including enzymes that act directly to attach or remove the ubiquitin moiety

Table 1 Factors regulating histone monoubiquitylation

Histone	Organism	Mono-ub residue	Role in transcription	E2	E3	Other factors	Ubiquitin proteases
H2A	Mouse	119	Gene silencing		*PRC1: **Ring1B**, <u>Mel18</u>, <u>Bmi1</u>, <u>Mpc2</u>, Mph1, Mph2, <u>Ring1A</u>		
	Drosophila	119	Gene silencing		*dPRC1: **dRing**, Pc, Ph, <u>Psc</u>		
	Human	119	Gene silencing		*hPRC1L: **Ring2**, <u>Ring1</u>, <u>Bmi1</u>, HPH2 <u>Mdm2</u>		Ubp-M
H2B	Yeast	123	Gene expression	Rad6	<u>Bre1</u>	*PAF: Paf1, Rtf1, Ctr9, Cdc73, Leo1 Kin28 BUR kinase Lge1	Ubp8 Ubp10
	Human	120	Gene activation	UbcH6	<u>RNF20/RNF40</u> <u>Mdm2</u>	*hPAF: Paf1, Ctr9, Cdc73, Leo1, Ski8	
	Drosophila	120					Usp7
H3	Mouse				<u>Np95</u>		
H4	Yeast	31	Transcription	Ubc4			
		91	termination	Ubc5			
	Human	31	Transcription	*EEUC:			
		91	termination	**Ubc5a,** Uba1, Uba3, Ubp5			
H1	Drosophila		Gene activation	*TFIID: **TAF$_{II}$250**			

Active factors (**bold type**) contained within complexes (*) are indicated, and the core proteins listed. RING domain proteins are underlined. For additional details and references, see text from histones and proteins that regulate the enzymatic machinery that promotes ubiquitylation. A review of these factors is presented in the following section and summarized in Table 1.

1.1
The Ubiquitin Conjugating Pathway

In this section, only a general overview of the ubiquitin conjugating pathway will be presented, as several excellent reviews of this topic have recently been published (Pickart 2001a; Pickart and Eddins 2004). Readers are directed to these reviews and the references therein for more detailed information on individual steps in this pathway. The first step in ubiquitin conjugation involves the ATP-dependent activation of ubiquitin by the enzyme Uba1 (E1), whereby ubiquitin becomes conjugated to the E1 through a covalent thioester linkage. Once activated, ubiquitin is transferred via a thioester bond to a cysteine residue in one of many different ubiquitin-conjugating enzymes (Ubcs/E2s). The ubiquitin moiety is transferred to the appropriate substrate through the intermediary of a large family of ubiquitin ligases (E3s), which contain several signature domains, including the RING and HECT domains (Joazeiro and Weissman 2000; Pickart 2001a). Ubiquitin is initially attached to the ε amino group of a specific lysine residue in a target protein through an isopeptide linkage involving ubiquitin's C-terminal glycine residue (G76). This monoubiquitin attachment can be built up into a polyubiquitin chain through the conjugation of additional ubiquitin molecules to each other. One of the most common polyubiquitin linkages is through lysine 48 of ubiquitin, which plays an essential role in targeting proteins for degradation (Pickart 1997; Pickart and Eddins 2004). Why proteins only become monoubiquitylated is not understood but might involve mechanisms that include steric hindrance or the exclusion of polyubiquitin chain catalyzing enzymes from complexes that mediate monoubiquitylation (Raasi and Pickart 2005).

The highly dynamic nature of histone ubiquitylation is underscored by the presence of multiple ubiquitin proteases (Ubps) in eukaryotic cells (16 in yeast and several hundred in vertebrates) (Hochstrasser 1995; Wilkinson 1997; Amerik, Li et al. 2000). These proteases act by hydrolyzing the linkage between ubiquitin and the target protein or between individual ubiquitin molecules in the case of polyubiquitylation. The large family of Ubps suggests that, like E3s, Ubps have significant specificity with respect their targets. Moreover, multiple Ubps may act on a single target. For example, as discussed below (Sect. 1.2.1), at least two different Ubps catalyze removal of ubiquitin from yeast H2B. This duplication of effort appears to target ub-H2B in different regions of chromatin for deubiquitylation, and may constitute a general mechanism for deubiquitylation of the same protein when it is present in different cellular compartments.

1.2
Factors Regulating Histone Ubquitylation

While a number of proteins have been implicated in either the attachment of removal of ubiquitin from histones, this review will focus only on those factors for which strong evidence exists for an enzymatic and/or biological function in histone ubquitylation or deubiquitylation. Table 1 contains a list of the factors with an established or highly likely role in the control of H2B, H2A, H4, H3, and H1 ubiquitylation. An emerging theme is that the monoubiquitylation of each histone is regulated by a different set of factors.

1.2.1
H2B

E2: H2B is monoubiquitylated on lysine 123 in yeast and on lysine 120 in other eukaryotes (Thorne, Sautiere et al. 1987; Robzyk, Recht et al. 2000). The E2 Rad6/Ubc2 targets yeast H2B for monoubiquitylation both in vitro and in vivo (Jentsch, McGrath et al. 1987; Sung, Prakash et al. 1988; Robzyk, Recht et al. 2000). Deletion of *RAD6* abolishes H2B ubiquitylation in both mitotic and meiotic yeast cells, suggesting that it encodes the sole Ubc in this organism that ubiquitylates H2B (Robzyk, Recht et al. 2000). Domain analysis of *RAD6* has identified two important regions for H2B ubiquitylation: an essential catalytic site (cysteine 88) and an acidic C terminal extension that appears to be required for optimal conjugation of ubiquitin to H2B (Morrison, Miller et al. 1988; Sung, Prakash et al. 1990; Robzyk, Recht et al. 2000; Sun and Allis 2002). Rad6 homologs have been identified in a wide variety of eukaryotic organisms, and it is a structurally conserved protein with the exception of an acidic C terminus that is unique to the *S. cerevisiae* enzyme (Sung, Prakash et al. 1988; Reynolds, Koken et al. 1990; Koken, Reynolds et al. 1991; Raboy and Kulka 1994; Wing and Jain 1995; Dor, Raboy et al. 1996; Koken, Hoogerbrugge et al. 1996; Roest, van Klaveren et al. 1996; Singh, Goel et al. 1998; Baarends, Hoogerbrugge et al. 1999; Roest, Baarends et al. 2004). In addition to its role in DNA damage repair (see below), Rad6 is also functionally conserved with respect to its role in catalyzing H2B ubiquitylation with the recent demonstration that a human Rad6 homolog, UbcH6, ubiquitylates vertebrate H2B in an in vitro system (Zhu, Zheng et al. 2005).

E3: Rad6 is a multifunctional ubiquitin-conjugating enzyme with several distinct cellular targets besides H2B, and each of these targets becomes ubiquitylated by Rad6's association with a different ubiquitin ligase (E3). Rad6 was originally identified as an enzyme with a role in postreplication repair of DNA damage (PRR) (Montelone, Prakash et al. 1981; Prakash 1981; Broomfield, Hryciw et al. 2001). One of its major targets in PRR is PCNA, which becomes polyubiquitylated through lysine 63 linkages of ubiquitin, a linkage that is not associated with protein turnover (Ulrich and Jentsch 2000; Hoege,

Pfander et al. 2002). It also targets short-lived proteins for polyubiquitylation through lysine 48 linkages, which leads to degradation by the proteosome (Watkins, Sung et al. 1993). The E3 Rad18 targets Rad6 to PCNA, while a separate ubiquitin ligase called Ubr1 directs Rad6 to ubiquitylate short-lived proteins (Bartel, Wunning et al. 1990; Bailly, Lauder et al. 1997; Xie and Varshavsky 1999; Hoege, Pfander et al. 2002; Stelter and Ulrich 2003; Ulrich 2004; Tasaki, Mulder et al. 2005). The Bre1 protein was identified as the E3 that directs Rad6 to monoubiquitylate yeast H2B in two different screens. A *bre1Δ* mutant was identified in a proteomics screen for factors that eliminated H3 lysine 4 (K4) methylation (which is regulated *in trans* by ub-H2B, Sect. 2), and also in a synthetic lethal screen for mutants that show inviability in combination with a deletion of the *HTZ1* gene, which encodes a histone H2A variant (Hwang, Venkatasubrahmanyam et al. 2003; Wood, Krogan et al. 2003). A *bre1Δ* mutant eliminates genome-wide H2B ubiquitylation in yeast cells, and the wild-type *BRE1* gene encodes a protein with a RING domain that is typically found in many E3 enzymes. The demonstration that Bre1 interacts with Rad6 and targets it to chromatin provides additional evidence that Bre1 is a *bona fide* E3 that directs Rad6 to its histone substrate (Wood, Krogan et al. 2003; Kao, Hillyer et al. 2004). Structural and functional homologs of Bre1 are found in other eukaryotes, but only in humans have any of the homologs been shown to regulate monoubiquitylation of H2B. Human homologs of yeast Bre1 are RNF20 and RNF40, two RING domain proteins that are $\sim 15\%$ identical and 28% similar to yeast Bre1 (Hwang, Venkatasubrahmanyam et al. 2003). RNF20/RNF40 were identified as H2B-specific ubiquitin ligases in a biochemical screen for factors that monoubiquitylate human H2B on lysine 120 (Zhu, Zheng et al. 2005). Like Bre1, RNF20/RNF40 interact with the human counterpart of Rad6, UbcH6, and the entire E2-E3 complex is required to ubiquitylate nucleosomal H2B in vitro. A Drosophila Bre1 homolog (dBre1) has also been identified (Bray, Musisi et al. 2005). Although it has not been shown to possess E3 ligase activity, it is a functional homolog of yeast Bre1 based on the defect in ub-H2B-regulated H3K4 methylation in dBre1 mutant cells.

Other factors promoting H2B ubiquitylation: a number of other factors have also been found to contribute to the regulation of H2B ubiquitylation in yeast. The majority of these factors have roles in transcription, particularly in transcription elongation, leading to the hypothesis that H2B ubiquitylation, like a number of other histone modifications, is co-transcriptionally regulated (Orphanides and Reinberg 2000; Gerber and Shilatifard 2003; Krogan, Kim et al. 2003; Ng, Robert et al. 2003; Xiao, Hall et al. 2003). A proteomics screen similar to the one that identified yeast Bre1 also revealed that components of the PAF transcription elongation complex are required for optimal ubiquitylation of H2B in this organism (Ng, Dole et al. 2003; Wood, Schneider et al. 2003; Xiao, Kao et al. 2005). While the precise function of PAF in the regulation of ub-H2B formation is not known, PAF interacts with Rad6 and is required for its association with elongating RNA polymerase II (Pol II) (Wood, Schneider et al. 2003; Xiao, Kao et al. 2005). Elongating Pol II, and specifically the Kin28 kinase that phosphorylates Pol II on serine 5 (Cismowski, Laff et al. 1995), is in turn globally required for H2B ubiquitylation through a mechanism that may involve the activation of Rad6 (Xiao, Kao et al. 2005). Another transcription elongation factor with a role in ub-H2B formation is the BUR kinase (Yao, Neiman et al. 2000). Mutations that reduce kinase activity simultaneously reduce the cellular levels of ub-H2B, an effect due in part to the reduced level of PAF association with genes in *bur* mutants and to the role of BUR in activating Rad6 by phosphorylation of serine 120 (Laribee, Krogan et al. 2005; Wood, Schneider et al. 2005). PAF is an evolutionarily conserved transcription elongation factor (Shi, Finkelstein et al. 1996; Shi, Chang et al. 1997; Chang, French-Cornay et al. 1999; Costa and Arndt 2000; Pokholok, Hannett et al. 2002; Squazzo, Costa et al. 2002; Kaplan, Holland et al. 2005; Zhu, Mandal et al. 2005), and recent reports suggest that its role in histone ubiquitylation is also functionally conserved. Human PAF interacts with UbcH6 and RNF20/RNF40, forming a trimeric complex that monoubiquitylates H2B in vitro (Zhu, Zheng et al. 2005). In addition, the PAF complex in *Arabadopsis* has a role in regulating the levels of ub-H2B-dependent H3K4 methylation, implicating it in the control of H2B ubiquitylation in plants (Oh, Zhang et al. 2004). Together, the data suggest that the following evolutionarily pathway controls the cellular levels of H2B ubiquitylation: (BUR) PAF → Elongating Pol II → Rad6/Bre1 → ub-H2B. Additional details of this pathway in the regulation of Ub-H2B formation in yeast are presented in Sect. 3.1.

Other factors have also been shown to regulate ub-H2b levels. In yeast, the product of the *LGE1* gene was identified in the same synthetic lethal screen that picked up *BRE1*, and an *lge1*Δ mutant shows globally reduced levels of cellular ub-H2B in this organism (Hwang, Venkatasubrahmanyam et al. 2003). Lge1 also interacts with Rad6 but does not contain a motif associated with typical E3 enzymes, and currently nothing is known about its function in regulating H2B ubiquitylation. The Mdm2 oncoprotein, a RING domain E3 ligase that targets p53 and acts at p53 regulated genes, also interacts with H2B and induces its monoubiquitylation in vitro and in vivo, albeit inefficiently (Minsky and Oren 2004). This raises the interesting possibility that "alternative" E3 ligases may regulate H2B ubiquitylation in a gene-specific context. Finally, yeast ub-H2B levels are globally regulated by glucose availability in cells (Dong and Xu 2004). H2B is deubiquitylated when glucose is depleted from the medium as cells enter stationary phase and rapidly (within minutes) reubiquitylated when glucose is added back to the medium. Although the mechanism underlying glucose-mediated ubiquitylation is not known, the induction of monoubiquitylation requires glycolysis, the central carbohydrate metabolic pathway. Thus, the ubiquitin conjugation machinery may be directly targeted by this pathway, which includes a number of protein kinases.

Ubps: several ubiquitin proteases targeting ub-H2B have been identified. The first was yeast Ubp8, which is a stoichiometric component of the SAGA histone acetyltransferase (HAT) complex (Henry, Wyce et al. 2003; Daniel, Torok et al. 2004; Powell, Weaver et al. 2004). Ubp8 specifically deubiquitylates H2B in transcriptionally active euchromatin through the targeted recruitment of SAGA to gene promoters (Henry, Wyce et al. 2003). As discussed below (Sect. 3.1), the antagonistic activities of Rad6-Bre1 and Ubp8 during transcription promote transient accumulation of ub-H2B at the promoters of the *GAL1* gene, which in turn is posited to control the balance between H3K4 and H3K36 (lysine 36) methylation, two marks associated with active transcription. A second yeast Ubp, Ubp10, also acts on ub-H2B, and both Ubp8 and Ubp10 have been shown to deubiquitylate Flag-ub-H2B in vitro and in vivo (Emre, Ingvarsdottir et al. 2005; Gardner, Nelson et al. 2005). In contrast to Ubp8, Ubp10 primarily targets ub-H2B present in transcriptionally silent, subtelomeric regions of the yeast genome that are adjacent to heterochromatin. However, Ubp10 may also have a role in deubiquitylating H2B in euchromatin as well, as a *ubp10Δ* mutant derepresses transcription of a number of genes that are not in or close to regions of heterochromatin (Gardner, Nelson et al. 2005). It has been postulated that the function of Ubp10 in heterochromatin is to maintain unmodified H2B, thereby preventing H3K4 and H3K79 (lysine 79) methylation in these regions, which can lead to the loss of transcriptional silencing (Gardner, Nelson et al. 2005). A similar scenario may occur at repressed euchromatic genes, which could potentially be activated by the ub-H2B-dependent methylation of H3K4. The essential Drosophila gene Usp7 also encodes a ubiquitin protease that selectively targets ub-H2B in an in vitro assay (van der Knaap, Kumar et al. 2005). A surprising feature of this deubiquitylating activity is that it is found in a stable complex with a metabolic enzyme, guanosine 5′-monophosphate synthetase (GMPS), which augments the ubiquitin protease activity of Usp7. Like Ubp10, Usp7 may help to maintain transcriptionally inactive chromatin by removing the potentially activating ubiquitin moiety from H2B (de Napoles, Mermoud et al. 2004). This is supported by the observation that Usp7 and GMPS contribute to epigenetic silencing of Polycomb group (PcG) regulated genes (van der Knaap, Kumar et al. 2005).

1.2.2
H2A

E2: H2A is ubiquitylated on lysine 119, except in yeast, where this modification has not been observed (Bohm, Crane-Robinson et al. 1980; Swerdlow, Schuster et al. 1990). The E2 that catalyzes monoubiquitylation of histone H2A is not known. While Rad6 is a likely candidate, data from mouse spermatids have shown that ub-H2A is present at wild-type levels when HR6B is knocked down (Baarends, Hoogerbrugge et al. 1999).

E3: ubiquitin ligases that specifically target H2A have been identified in both vertebrates and flies as components of the Polycomb group (PcG) complex PRC1, which is involved in epigenetic gene silencing (de Napoles, Mermoud et al. 2004; Fang, Chen et al. 2004; Wang, Wang et al. 2004; Zhang, Cao et al. 2004). These ligases are RING domain-containing proteins (mouse Ring1B; human Ring2; fly dRing), and the RING domain is essential for ubiquitin ligase activity in vitro. PRC1 and ub-H2A are enriched at regions of the mouse and fly genomes that are subject to PcG-dependent transcriptional silencing, such as the inactive X chromosome (Xi) and homeotic genes, and the presence of the RING E3 is required for H2A ubiquitylation at these locations (de Napoles, Mermoud et al. 2004; Fang, Chen et al. 2004; Wang, Wang et al. 2004). Interestingly, PRC1 contains two additional RING domain-containing proteins, Ring1A and Bmi1, but, unlike Ring1B, neither possesses a functional ubiquitin ligase activity (Y. Zhang, personal communication). However, the two proteins stimulate the ligase activity of Ring1B in the PRC1 complex in vitro and control the levels of ub-H2A at silenced Hox genes in vivo (Cao, Tsukuda, Zhang 2005). The oncoprotein Mdm2 has also been reported to promote monoubiquitylation of H2A as well as H2B (Minsky and Oren 2004) and, as discussed above, this activity may be restricted to genes regulated by p53. In contrast, Ring1B apparently regulates the levels of H2A ubiquitylation on a global level (de Napoles, Mermoud et al. 2004).

Ubps: the only strong candidate for an H2A deubiquitylating activity is human Ubp-M, which has been reported to cleave ubiquitin from H2A in vitro (Cai, Babbitt et al. 1999). This enzyme associates with mitotic chromosomes at the same time as histones are globally deubiquitylated during metaphase, and is postulated to promote ubiquitin cleavage from H2A at this point in the cell cycle. Ub-H2A is also deubiquitylated during apoptosis, and Ubp-M may also control cleavage during this event as well (Mimnaugh, Kayastha et al. 2001).

1.2.3
H4

E2: although H4 was not thought to be ubiquitylated, it was recently found that lysines 31 and 91 of H4 are monoubiquitylated in both human and yeast cells (D. Reinberg, personal communication). A monoubiquitin conjugating activity specific for H4 resides in a human complex called EEUC (E1-, E2-, Ubp-containing complex). This complex contains conserved activities that include the Uba1 ubiquitin activating enzyme, the E2 UbcH5a, and two ubiquitin proteases (D. Reinberg, personal communication). UbcH5a is required for optimal ubiquitylation activity of EEUC both in vitro and in vivo, and its yeast homologs Ubc4 and Ubc5 appear to promote ubiquitin conjugation to H4 globally in vivo.

E3: no ubiquitin ligase activity targeting H4 has been identified.

Ubps: two Ubps (Ubp3, Ubp5) are subunits of the EEUC complex and important for the transcriptional activity of EEUC (see Sect. 4), but their activity in the removal of ubiquitin from ub-H4 has not been directly tested (D. Reinberg, personal communication).

1.2.4
H1

Histone H1 has been reported to be monoubiquitylated by the multifunctional Drosophila $TAF_{II}250$, which is the largest subunit of the $TBP-TAF_{II}$ complex (Ruppert, Wang et al. 1993; O'Brien and Tjian 2000; Pham and Sauer 2000). The residue(s) targeted for ubiquitylation are not known. In vitro studies indicate that $TAF_{II}250$ has both a ubiquitin activating activity (E1) and a ubiquitin conjugating activity (E2) that are specific for H1, with these activities residing in the $TAF_{II}250$ C terminus (Pham and Sauer 2000). Point mutations in this domain (called ubac for the ubiquitin activating/conjugating domain) reduce ub-H1 levels in fly embryos and impair expression of several mesoderm-determining genes (Pham and Sauer 2000).

1.2.5
H3

Histone H3 has been shown to be ubiquitylated in rat spermatids, although the site of ubiquitylation is unknown (Chen, Sun et al. 1998). The RING domain-containing murine protein Np95 plays an important role in cell-cycle progression and exhibits ubiquitin ligase activity towards the core histones in vitro, with a strong preference for the N terminal tail of H3 (Citterio, Papait et al. 2004).

2
Relationship Between Histone H2B Ubiquitylation and Histone H3 Methylation

One of the most exciting findings of the past several years was the discovery that ubiquitylation of H2B in yeast is required for the unidirectional methylation of histone H3 on lysines 4 and 79 (H3K4me and H3K79me) (Briggs, Xiao et al. 2002; Dover, Schneider et al. 2002; Ng, Xu et al. 2002; Sun and Allis 2002). Recent data suggest that this represents an evolutionarily conserved phenomenon that is likely to play an important role in establishing the proper patterns of H3 methylation over genes during transcription. The relationship between ub-H2B and H3K4me/K79me differs from other forms of cross-talk between histone modifications in several important respects. First, it represents regulation *in trans* rather than *in cis* (an example of which is phosphorylation of H3 on serine 10 regulating acetylation of H3 on lysine 14) (Fischle, Wang et al. 2003; Lo, Henry et al. 2004). Second, the effect is exerted on a genome-wide level, with H3K4me and H3K79me levels globally reduced in the absence of ub-H2B. H3K4 and H3K79 are mono-, di-, and trimethylated and, while it was initially proposed that ub-H2B controls all three methylation states, recent data indicate that it preferentially regulates the di- and trimethylated forms of the two H3 lysine residues (Shahbazian, Zhang et al. 2005).

Two key issues are how ub-H2B regulates H3K4/K79 di- and trimethylation and what are the biological consequences of this trans-histone regulatory pathway. Both H2B ubiquitylation and H3K4 methylation are in large part controlled by a co-transcriptional mechanism, in which the enzymes that catalyze each modification associate with elongating RNA polymerase II through the transcription elongation factor PAF (see Sect. 3.1.1 below) (Krogan, Dover et al. 2003; Ng, Robert et al. 2003; Wood, Schneider et al. 2003; Xiao, Kao et al. 2005). Ub-H2B is apparently not required for the chromatin association of enzymes that directly catalyze H3K4 and H3K79 methylation (Set1 and Dot1), but is required for the recruitment of proteosomal subunits to actively transcribed genes, and these subunits, in turn, play an undefined role in H3K4 and H3K79 dimethylation (Ng, Robert et al. 2003; Ezhkova and Tansey 2004). Mutations in the BUR kinase, a putative transcription elongation factor, reduce ub-H2B levels, in part by inhibiting recruitment of PAF to gene-coding regions, and selectively eliminate H3K4 trimethylation (Yao, Neiman et al. 2000; Laribee, Krogan et al. 2005). This suggests that the absolute levels of ub-H2B in chromatin could be a major determinant as to whether two or three methyl residues will be attached to lysine 4, which can be monoubiquitylated in the absence of ub-H2B (Shahbazian, Zhang et al. 2005). These data have led to several possible models for regulation of H3K4 and K79 methylation by ub-H2B. Two related models posit that the presence of the bulky ubiquitin moiety on nucleosomal H2B alters chromatin structure locally (either at the level of individual nucleosomes or over a large domain) to permit Set1 and Dot1 access to their substrates (Henry and Berger 2002; Zhang 2003; Xiao, Kao et al. 2005). In contrast to this "structural" model, a second model suggests that the ubiquitin moiety on H2B provides an interaction surface for factors that regulate K4 and K79 methylation. This surface could help to assemble complexes that promote di- and trimethylation or, by binding the histone methyltransferases (HMTs) themselves, influence their catalytic activity. At this point in time, there is no evidence that specifically favors either model. In vitro reconstitution studies have not revealed differences in the structural properties of nucleosomes that contain ubiquitylated histones (Davies and Lindsey 1994; Jason, Moore et al. 2001, 2002; Moore, Jason et al. 2002). In addition, ubiquitin binding has not been reported for factors that regulate H3K4 and H3K79 methylation. However, it was recently reported that the H3K79 HMT, Dot1, contains two UBA-like domains that are present in numerous monoubiquitin-binding proteins involved in intracellular trafficking, suggesting that Dot1 might bind the ubiquitin moiety on H2B (Shahbazian, Zhang et al. 2005). How ub-H2B regulates trans-histone methylation thus remains a challenging and fascinating area of histone research.

3
Role of Histone Ubiquitylation in Gene Expression

What we currently know about the roles of ubiquitylated histones in the regulation of transcription comes almost exclusively from studies of ub-H2B and ub-H2A. These two species of ubiquitylated histones were initially found to be enriched in nucleosomes at actively transcribed genes (Levinger and Varshavsky 1982; Barsoum and Varshavsky 1985; Nickel, Allis et al. 1989; Davie, Lin et al. 1991), and for many years it was believed that this enrichment represented a primary role in transcriptional activation. However, recent data suggest that histone ubiquitylation, like histone methylation, may play both positive and negative roles in transcription, and an emerging theme is that ub-H2B is associated with transcriptional activation and ub-H2A with transcriptional silencing. The following sections will review recent data that have helped to define the roles of ub-H2B and ub-H2A in these transcriptional processes. In addition, recent data linking H4 ubiquitylation, a newly identified modification, to the regulation of transcription termination will also be discussed.

3.1
Ubiquitylated H2B

The discovery of ub-H2B in yeast made it possible to study the cellular roles of this histone modification from a genetic perspective through analysis of mutations in the H2B residue that becomes ubiquitylated (K123) or in the genes encoding components of the ubiquitylation machinery. In addition, construction of special strains that can be used to measure the levels of ub-H2B in chromatin by sequential double chromatin immunoprecipitation (ChDIP) of Flag-H2B and HA-ubiquitin has revealed the dynamic nature of this histone modification during transcription initiation and elongation, as well as a partial picture of the distribution of ub-H2B in the yeast genome (Henry, Wyce et al. 2003). ChDIP, for example, has shown that ub-H2B is present in transcriptionally active chromatin and at greatly reduced levels in transcriptionally silent chromatin such as telomere-associated regions and the silent mating type loci (Kao, Hillyer et al. 2004; Emre, Ingvarsdottir et al. 2005).

3.1.1
Gene Activation

Evidence that ub-H2B has a role in activated transcription in yeast first came from the analysis of an *htb1-K123R* mutant, in which the site of ubiquitin conjugation (K123) was changed to a residue (R123) that cannot be ubiquitylated (Henry, Wyce et al. 2003; Kao, Hillyer et al. 2004). This mutant grows poorly on media containing raffinose or galactose compared to a wild-type strain. These carbon sources induce transcription of the *SUC2* and *GAL* genes, respectively, both of which are highly regulated in response to the appropriate environmental stimulus. In an *htb1-K123R* mutant, *SUC2*, *GAL1*, and *PHO5* mRNAs accumulate to ~ 50% of wild-type levels and there is a delay in the appearance of all three RNAs. Moreover, a synthetic slow growth phenotype and a severe defect in *SUC2*, *GAL1*, and *PHO5* gene expression occur when an *htb1-K123R* mutation is combined with mutations in genes that encode subunits of the Swi/Snf or SAGA complexes, which represent transcriptional co-activators that remodel or modify chromatin. Together, the combined data indicate that ub-H2B contributes to activated transcription through its presence in chromatin, and that this function overlaps with activities that also act on chromatin to induce transcription.

The role of ub-H2B in activated transcription is exerted at the level of both transcription initiation and transcription elongation. In addition to promoting slow growth on galactose- or raffinose-containing medium, mutations affecting H2B ubiquitylation (e.g. *htb1-K123R*, *rad6Δ*, *bre1Δ*) also confer heightened sensitivity to the drug 6-azauracil (6-AU), a drug that causes "elongation stress", and result in synthetic lethality or slow-growth phenotypes when combined with mutations in genes encoding transcription elongation factors (Xiao, Kao et al. 2005). As discussed below, one effect of the absence of ub-H2B is a decrease in the levels of RNA polymerase II (Pol II) over the coding region of the activated *GAL1* gene during the initial stages of gene induction. This could represent a decrease in the rate of Pol II initiation, slower entry of Pol II into the coding region, stability of Pol II over the ORF, or some combination of these effects.

Detailed studies of the role of ub-H2B in gene activation have been performed primarily on the *GAL1* gene, and have led to the following picture of the events occurring during the interrelated processes of transcription initiation and elongation (Xiao, Kao et al. 2005). When *GAL1* transcription is induced by shifting cells from glucose- (repressing) or raffinose- (noninducing) containing medium to galactose- (activating) containing medium, one of the first factors to be recruited to the upstream activating site (UAS) is Rad6 (Kao, Hillyer et al. 2004). Rad6 recruitment depends on both the Gal4 activator and the E3 ligase Bre1, and leads to ubiquitylation of H2B on the nucleosome that covers the *GAL1* TATA element (Henry, Wyce et al. 2003; Wood, Krogan et al. 2003; Kao, Hillyer et al. 2004). Ubiquitylation of H2B is closely followed by recruitment of the multisubunit SAGA co-activator to the UAS (Kao, Hillyer et al. 2004). SAGA recruitment serves multiple important roles in the initiation of *GAL1* transcription: the Spt3 subunit recruits the TATA binding factor TBP; it regulates recruitment of Mediator components Srb8-Srb11; and a SAGA module that includes the Ubp8 ubiquitin protease removes ubiquitin from H2B (Dudley, Rougeulle et al. 1999; Sterner, Grant et al. 1999; Larschan and Winston 2001; Bhaumik and Green 2002; Ingvarsdottir, Krogan et al. 2005; Larschan and Winston 2005; Lee, Florens et al. 2005). One of the surprising aspects of *GAL1* activation is that efficient initiation of transcription requires sequential ubiquitylation and deubiquitylation of H2B, which was first revealed by the similar delay in *GAL1* mRNA accumulation in *htb1-K123R* and *ubp8Δ* mutants (Henry, Wyce et al. 2003). Ubiquitylation of H2B coincides with the recruitment of Pol II, and in the absence of ub-H2B Pol II recruitment is significantly delayed, suggesting an important role for ubiquitin attachment in this event (Xiao, Kao et al. 2005; C.-F. Kao, unpublished data). While it is not fully understood why initiation should also depend on the subsequent removal of ubiquitin from H2B, one clue to this puzzle is that in the absence of Ubp8-mediated deubiquitylation, there is an imbalance in the methylation states of lysine 4 and lysine 36 on histone H3 at the *GAL1* core promoter: H3K4me2/me3 levels rise and H3K36me2 levels decrease compared to wild-type cells (Henry, Wyce et al. 2003). Just the opposite scenario is seen in the absence of ub-H2B, where H3K4me2/me3 are abolished and H3K36me levels rise. Both sets of methyl marks have been associated with active transcription, with H3K4me3 concentrated at the promoter and 5′ORF of many expressed genes and H3K36me2 present closer to the 3′ end of coding regions (Bernstein, Humphrey et al. 2002; Santos-Rosa, Schneider et al. 2002; Ng, Robert et al. 2003; Bannister, Schneider et al. 2005; Pokholok, Harbison et al. 2005; Rao, Shibata et al. 2005). Thus, shifts in the levels or distribution of these marks could lead to downstream effects on the recruitment or distribution of factors with roles in transcription initiation and elongation (Henry, Wyce et al. 2003; Zhang 2003; Emre and Berger 2004).

As Pol II begins to traverse the ORF, Rad6 leaves the UAS and spreads throughout the coding region, essentially piggybacking on Pol II (Kao, Hillyer et al. 2004; Xiao, Kao et al. 2005). Rad6 association with Pol II requires at least two factors – Bre1 and the PAF transcription elongation complex (Wood, Schneider et al. 2003; Xiao, Kao et al. 2005). Bre1 thus plays a role at two points: it is required for Rad6's initial recruitment to the *GAL1* promoter and again for Rad6's interaction with Pol II in the coding region. In contrast, the PAF complex, which is localized to gene promoters, coding regions, and 3'ends (Krogan, Kim et al. 2002; Pokholok, Hannett et al. 2002; Simic, Lindstrom et al. 2003), only mediates Rad6's association with Pol II. As predicted from the coding region association of Rad6, ub-H2B is also present throughout the *GAL1* ORF, as well as the ORFs of a number of constitutively expressed genes (Xiao, Kao et al. 2005). The presence of ub-H2B at the ORF requires not only Rad6, but also Bre1, the PAF complex, and Kin28, the Pol II CTD-serine 5 kinase (Xiao, Kao et al. 2005). While the requirement for Rad6 and Bre1 is obvious based on their biochemical activities in ubiquitin conjugation, the requirement for PAF and Kin28 is still mysterious. Several related scenarios can be envisioned based on the observation that PAF itself associates with Pol II (Krogan, Kim et al. 2002; Pokholok, Hannett et al. 2002; Squazzo, Costa et al. 2002). The PAF complex could contain an activity that activates Rad6-Bre1, but this activity would be stimulated only when PAF is associated with Pol II phosphorylated on serine 5 of the CTD, in which case the requirement for Kin28 might be indirect. Alternatively, Ser 5 phosphorylation of the CTD could recruit another factor that activates Rad6-Bre1, but activation would occur only when the ubiquitylation machinery is tethered to Pol II via PAF, in which case the requirement for PAF would be indirect. Thus, how Rad6-Bre1 is co-transcriptionally activated remains a key issue to be resolved. An intriguing possibility is that PAF and the BUR kinase performutually reinforcing functions activate Rad6.

Rad6 and ub-H2B each turn over during both the initiation and elongation phases of *GAL1* transcription (Henry, Wyce et al. 2003; Kao, Hillyer et al. 2004; Xiao, Kao et al. 2005). This pattern suggests that H2B ubiquitylation might define a pioneer round of transcription that marks a gene as having just been activated or constitute a mark of transcriptional "memory" through its regulation of H3K4 methylation (Orphanides and Reinberg 2000; Ng, Robert et al. 2003). Ubp8-dependent deubiquitylation of H2B occurs at the *GAL1* promoter through the recruitment of SAGA, and recent data indicate that Ubp8, by associating with elongating Pol II, is also present over the *GAL1* coding region (Henry, Wyce et al. 2003; S. Berger and B. Strahl, personal communication). As discussed above, this turnover could ultimately limit the coding region levels or distribution of H3K4 methylation by Set1, which, like Rad6, travels with elongating Pol II (Ng, Robert et al. 2003).

Very little is known about the role that ub-H2B plays in either the initiation or the elongation phases of transcription. As discussed above (Sect. 2), this role could be structural, with the bulky ubiquitin moiety helping to open up chromatin, thus allowing recruitment or access of factors with roles in these processes. Alternatively, ubiquitin could serve as a binding site for assembly of initiation or elongation complexes or for activation of enzymatic activities that reside in initiation or elongation factors. A complicating factor is the trans-histone regulation of H3K4me2/me3 and H3K79me2/me3 by ub-H2B, i.e. some of the transcriptional roles ascribed to ub-H2B may be unique to this modified histone while others may be due to downstream effects on H3 methylation. H3K4me2/me3, in particular, have been associated with transcriptionally active chromatin (Bernstein, Humphrey et al. 2002; Kouzarides 2002; Santos-Rosa, Schneider et al. 2002; Pokholok, Harbison et al. 2005), and H3K4me is known to bind Chd1, a chromodomain protein with roles in transcription elongation (Simic, Lindstrom et al. 2003; Pray-Grant, Daniel et al. 2005). However, other evidence points to a unique role for ub-H2B in transcription elongation. Mutations in the H2B ubiquitylation machinery (*rad6Δ, bre1Δ, htb1-K123R*) confer 6-AU sensitivity and exhibit genetic interactions with mutations in genes encoding proteins with roles in transcription elongation (Xiao, Kao et al. 2005). These phenotypes are not seen in the absence of H3K4 or H3K79 methylation (Xiao, Kao et al. 2005). How ub-H2B affects transcription initiation and elongation thus remains a highly important area of research.

The role of ub-H2B in transcription in human cells has been revealed only recently. As outlined in Sect. 1 above, there is significant conservation between the yeast and human H2B ubiquitylation systems, with Rad6-Bre1 and PAF contributing to the monubiquitylation of H2B in both organisms. Not only do a similar group of factors monoubiquitylate human H2B, but the trans-histone regulation of H3K4 and H3K79 methylation by ub-H2B is also conserved in human cells (Zhu, Zheng et al. 2005). As in yeast, H2B ubiquitylation in humans also appears to be a co-transcriptional event, with the entire ubiquitylation machinery (Rad6-Bre1-PAF) being recruited to transcriptionally active genes in vivo and spreading across coding regions (Zhu, Zheng et al. 2005). Finally, human ub-H2B has also been linked to gene activation, and specifically to activation of the developmentally important *Hox* genes (Zhu, Zheng et al. 2005). This connection may be through the trans-histone regulation of H3K4 methylation, which is controlled by the Set1-containing MLL complex and known to be associated with *Hox* gene expression (Milne, Briggs et al. 2002; Dou, Milne et al. 2005; Wysocka, Swigut et al. 2005).

3.1.2
Gene Silencing

Although ub-H2B is strongly connected to gene activation, other data indicate that it also plays a role in gene silencing in yeast. Mutations that abolish H2B ubiquitylation (e.g. *rad6Δ, bre1Δ, htb1-K123R*) decrease silencing of telomere-associated genes, whereas deletion of *UBP10*, which encodes a ubiquitin protease targeting ub-H2B present in telomere-proximal chromatin, strengthens silencing (Huang, Kahana et al. 1997; Singh, Goel et al. 1998; Dover, Schneider et al. 2002; Sun and Allis 2002; Wood, Krogan et al. 2003; Emre, Ingvarsdottir et al. 2005; Gardner, Nelson et al. 2005). The role of ub-H2B in gene silencing appears to be indirect and closely related to its regulation of H3K4 and H3K79 methylation. Silencing in yeast is dependent on a complex of Sir proteins (Sir2/Sir3/Sir4) that are localized to silent chromatin through their interactions with underacetylated and undermethylated N terminal tails of histones H3 and H4 (Nislow, Ray et al. 1997; Bryk, Briggs et al. 2002; Ng, Feng et al. 2002; van Leeuwen, Gafken et al. 2002; Ng,

Ciccone et al. 2003; Rusche, Kirchmaier et al. 2003; Orlandi, Bettiga et al. 2004; Schneider, Wood et al. 2005). In the absence of H2B ubiquitylation, the precipitous drop in the levels of H3K4me2/me3 and H3K79me2/me3 in euchromatin acts as a sink for telomere-associated Sir proteins, which spread from silent chromatin and result in weakened silencing (van Leeuwen, Gafken et al. 2002; van Leeuwen and Gottschling 2002). Conversely, the Ubp10-dependent maintenance of ub-H2B and, in turn, H3K4 and H3 K79 methylation in telomere-proximal regions acts as a strong buffer to the loss of Sir proteins from heterochromatin (Emre, Ingvarsdottir et al. 2005; Gardner, Nelson et al. 2005).

H2B ubiquitylation has also been linked to repression of some euchromatic yeast genes. This was first revealed by analysis of the regulation of the *ARG1* gene, which is repressed by the ArgR/Mcm1 repressor complex in a sequence-specific manner in the presence of exogenous arginine (Amar, Messenguy et al. 2000). Mutations that eliminate H2B ubiquitylation (*rad6Δ*, *bre1Δ*, *htb1-K123R*) derepress *ARG1* in arginine-containing medium without significantly affecting activated transcription (Turner, Ricci et al. 2002). Ub-H2B dependent *ARG1* repression requires the ArgR/Mcm1 repressor, but it is not known if this effect is mediated through the trans-histone regulation of H3K4 or H3K79 methylation. Interestingly, SAGA is also required for *ARG1* repression and appears to function in the same pathway as H2B ubiquitylation (Ricci, Genereaux et al. 2002; Turner, Ricci et al. 2002). Increases in the levels of ub-H2B in a *ubp8Δ* mutant strengthen *ARG1* repression (Lee, Florens et al. 2005), raising the intriguing possibility that the ubiquitylation and deubiquitylation of H2B play a role in both repression and activation of a subset of yeast genes. Basal repression of the *PHO5* and *GAL1-10* genes is also partially relieved in the absence of regulators of H2B ubiquitylation (*rad6Δ*, *bre1Δ*, *lge1Δ*, *paf1Δ*, *rtf1Δ*) (Carvin and Kladde 2004). However, it is likely that these latter effects are due to the absence of H3K4 methylation as *set1Δ* mutations confer a similar phenotype. Finally, transcriptional profiling with mutants defective in the control of H2B deubiquitylation (*ubp8Δ*, *ubp10Δ*) have revealed sets of euchromatic genes that are derepressed by the presence of elevated levels of ub-H2B (Gardner, Nelson et al. 2005). Loss of Ubp10 increases the steady-state levels not only of ubiquitylated H2B but also of H3K4 di- and trimethylation, which, in turn, are associated with active transcription (Emre, Ingvarsdottir et al. 2005; Gardner, Nelson et al. 2005). It has been suggested that Ubp10 acts at repressed genes to deubiquitylate H2B and thereby limit the amount of activating H3K4 methylation. Overall, the general picture that emerges is that the role of ub-H2B in gene silencing is probably indirect and a consequence of the downstream effects of ub-H2B on the global or gene-specific levels of H3K4 and/or H3K79 methylation. Methylation could eliminate binding sites for silencing or repressor complexes such as observed with the Sir repressors or provide novel binding sites for other repressor complexes such as ArgR/Mcm1.

3.2
Ubiquitylated H2A

Like ub-H2B, ub-H2A was long thought to play only an activating role in gene expression. This view was challenged by the finding that ub-H2A could be found in both active and inactive regions of chromatin, and it is now recognized that ub-H2A is an epigenetic mark associated with transcriptional silencing. A key reagent in the analysis of ub-H2A was the development of a monoclonal antibody that specifically recognizes this H2A modification in both immunofluorescence and chromatin immunoprecipitation assays (Vassilev, Rasmussen et al. 1995). One of the first clues that ub-H2A was connected to transcriptionally silent chromatin came from cytological analysis of mouse spermatocytes, which showed that ub-H2A was present in meiotic prophase cells in the sex body or XY body, a region of silent heterochromatin (Baarends, Hoogerbrugge et al. 1999). Additional cytological studies have revealed that ub-H2A is also present in other regions of heterochromatin such as the inactive X chromosome (Xi) of female mammals and unpaired autosomal regions in male meioses and unpaired X and Y chromosomes in female meioses (Smith, Byron et al. 2004; Baarends, Wassenaar et al. 2005). However, although ub-H2A is concentrated in these heterochromatic regions, it is also present globally in euchromatin.

As outlined in Table 1 and discussed in Sect. 1, the E3 that directs ubiquitin attachment to H2A has been identified as Ring1B (hRing2/dRing), a RING domain protein that is a component of the Polycomb group (PcG) complex PRC1, a factor implicated in heritable gene silencing in flies and vertebrates (de Napoles, Mermoud et al. 2004; Fang, Chen et al. 2004; Wang, Wang et al. 2004; Zhang, Cao et al. 2004; Dejardin and Cavalli 2005). Several lines of evidence have linked these Ring1B homologs and ub-H2A to various forms of gene silencing. Cytological studies have shown that PRC1 and ub-H2A are simultaneously enriched on the inactive X chromosome early in the process of X inactivation in the mouse (de Napoles, Mermoud et al. 2004; Fang, Chen et al. 2004). X inactivation is characterized by an initiation phase, in which chromatin on the Xi chromosome becomes silenced, and a maintenance phase in which the silent chromatin established on Xi remains stable during somatic cell divisions (Jaenisch, Beard et al. 1998; Cohen, Royce-Tolland et al. 2005). Interestingly, the Xi enrichment of PRC1 is transient, suggestive of a role for the H2A modification in the initiation phase (de Napoles, Mermoud et al. 2004; Fang, Chen et al. 2004). However, there are conflicting data on whether ub-H2A itself is transient, so it is unclear whether ub-H2A acts later during the maintenance phase or plays a role during both phases of X inactivation (de Napoles, Mermoud et al. 2004; Fang, Chen et al. 2004). In flies, ChIP analysis has demonstrated that dRing and ub-H2A co-localize to the PRE (Polycomb response element) that is the binding site for PcG complexes involved in silencing of the homeotic gene *Ubx* (Wang, Wang et al.

2004; Wang, Brown et al. 2004; Zhang, Cao et al. 2004; Dejardin and Cavalli 2005). A similar scenario occurs in humans, where ChIP has shown that PRC1 subunits and ub-H2A are locally present at the promoter of the transcriptionally silenced *HoxC13* gene (Cao, Tsukuda, Zhang 2005). Finally, ub-H2A is enriched on unpaired, silenced chromosomes present in male and female meioses, and the timing of its association with these unpaired regions suggests that it might play a role in the maintenance of gene silencing (Baarends, Wassenaar et al. 2005).

The strongest evidence that PRC1 complexes and ub-H2A play direct roles in the control of transcriptional silencing comes from RNAi experiments directed against the RING E3s. RNAi-mediated knockdown of dRing in flies leads to derepression of the homeotic gene *Ubx* (Wang, Wang et al. 2004). Similarly, human cells deficient for Bmi1, which regulates Ring1B's ligase activity, show derepressed transcription of a number of *Hox* genes (Cao, Tsukuda, Zhang 2005). In contrast, depletion of ub-H2A on Xi by RNAi-mediated knockdown of Ring1B and its homolog Ring1A does not lead to reactivation of Xi-linked genes in the mouse (de Napoles, Mermoud et al. 2004). As mentioned above, while ub-H2A is enriched in many regions of inactive chromatin, it is also present globally throughout the mouse genome (de Napoles, Mermoud et al. 2004). Whether these regions of ub-H2A enrichment represent inactive regions of chromatin is not known; however, the genome-wide association of ub-H2A is also dependent on Ring1B (de Napoles, Mermoud et al. 2004). This suggests that ub-H2A is primarily present at regions controlled by PRC1. Although *Hox* genes are one of the primary targets of PRC1 complexes, the full spectrum of genes directly controlled by PRC1 is not known. This suggests that PRC1 and ub-H2A may have roles in the regulation of a wide variety of mammalian genes.

Almost nothing is known about how ub-H2A regulates transcriptional silencing. Ub-H2A associates predominantly with *Hox* gene promoters (Cao, Tsukuda, Zhang 2005), suggesting that it could exert a local effect on chromatin structure or on the recruitment of other repressive factors that inhibit transcription initiation. A second PcG complex called PRC2 is also recruited to inactive X genes in mammals and co-localizes with PRC1 at fly and vertebrate *Hox* genes (Zhang, Cao et al. 2004). PRC2 contains an HMT that catalyzes methylation of H3 on lysine 27 (H3K27me), a mark essential for silencing of the inactive X and homeotic genes (Cao, Wang et al. 2002; Muller, Hart et al. 2002; Plath, Fang et al. 2003). H3K27 methylation plays a role in PRC1 recruitment to these regions; however, H3K27 methylation itself is not affected in PRC1 knockdown cells that are deficient in ub-H2A, indicating that ub-H2A does not affect silencing through the trans-histone regulation of this particular methyl mark (de Napoles, Mermoud et al. 2004; Wang, Wang et al. 2004; Cao, Tsukuda, Zhang 2005). Other scenarios for ub-H2A function in gene silencing include regulation of other histone modifications associated with PcG-dependent silencing or antagonism of activating modifications; effects on the activity of histone deacetylases (HDACs) that promote silencing; regulation of the deposition of the variant histone macroH2A to regions of silent chromatin; or regulation of the association of the linker histone H1 with regions of chromatin targeted for compaction (Chadwick and Willard 2003; Francis, Kingston et al. 2004; Zhang, Cao et al. 2004; Hernandez-Munoz, Lund et al. 2005; Jason, Finn et al. 2005). Finally, ub-H2A might play a role in the PcG-mediated antagonism of chromatin remodeling by trithorax group (Trx) complexes, which play roles in gene activation through ATP-dependent nucleosome remodeling (Hanson, Hess et al. 1999; Francis, Saurin et al. 2001). Clearly, this is an exciting area of research that should yield much new information about the role of ub-H2A in the epigenetic regulation of gene silencing over the next several years.

3.3
Ubiquitylated H4

The newly discovered ubiquitylation of H4 has been connected to transcription termination through analysis of the EEUC complex (Table 1) that mediates ubiquitin attachment to this histone. Synthetic genetic array (SGA) analysis in yeast, in which EEUC complex mutants were crossed to a library of ~ 400 yeast deletion strains defective in some aspect of transcription or chromatin metabolism, revealed genetic interactions between EEUC proteins Ubc4 and Ubp3 and proteins implicated in transcription termination (D. Reinberg, personal communication). *ubc4Δ* and *ubp3Δ* mutants show transcription terminator read-through in vivo, and EEUC components ChIP across the coding region and polyadenylation site on transcriptionally active yeast and human genes (D. Reinberg, personal communication). Together, the genetic and functional data are consistent with a role for EEUC in efficient transcription termination in vivo. The finding that a ubiquitin conjugating enzyme (Ubc4) and a deubiquitylation enzyme (Ubp3) are physically associated and functionally equivalent raises the intriguing possibility that the attachment and removal of ubiquitin from histones represents a conserved paradigm for the control of various steps in activated transcription. Although H4 lysine residues 31 and 91 are monoubiquitylated by EEUC, only ubiquitylation of lysine 91 is important for transcription termination. Interestingly, this same residue is also acetylated by the nuclear Hat1-Hat2-Hif1 complex, and K91 *acetylation* has been linked to the repair of DNA damage in yeast (Ye, Ai et al. 2005). Lysine 91 lies in a region of H4 important for the interaction between the H3/H4 tetramer and H2A/H2B dimers, and it has been suggested that acetylation of this site might influence the formation or stability of the histone octamer. Targeted ubiquitylation of H4 on K91 during transcription elongation could also create a chromatin structure permissive for the access of termination factors, or serve as a platform to recruit such factors.

4
Additional Cellular Roles of Ubiquitylated Histones

While this review has focused on the connection between ubiquitylated histones and gene expression, histone ubiquitylation is very likely to have roles in other chromosomal processes as well. One role that has recently emerged is in the control of DNA double-strand breaks (DSBs). Mutations that eliminate ub-H2B in yeast (*rad6Δ, bre1Δ, htb-K123R*) cause a meiotic defect that results from a decrease in the formation of DSBs by the enzyme Spo11 (Yamashita, Shinohara et al. 2004). Yeast ub-H2B also plays a role in the cellular response to DNA damage: the failure to form ub-H2B after DSB induction induces the DNA damage signaling pathway and causes a checkpoint defect characterized by the absence of cell-cycle arrest (San-Segundo and Roeder 2000; Game, Williamson et al. 2005; Giannattasio, Lazzaro et al. 2005; Wysocki, Javaheri et al. 2005). As in transcription, the role of ub-H2B in DSB formation and the DNA damage response may ultimately be mediated through its trans-histone regulation of H3K4 and H3K79 methylation. Mutations that specifically eliminate H3K4 methylation lead to a meiotic defect similar to that observed in the absence of ub-H2B, while the checkpoint deficiency of mutants in the H2B ubiquitylation pathway is due primarily to a defect in H3K79 methylation (Sollier, Lin et al. 2004; Giannattasio, Lazzaro et al. 2005; Wysocki, Javaheri et al. 2005). In the case of H3K79 methylation, the methyl mark appears to act in part as a binding platform for the recruitment of proteins that mediate the checkpoint response to DNA damage (Huyen, Zgheib et al. 2004; Wysocki, Javaheri et al. 2005). Thus, as ubiquitylated histones undergo more scrutiny, it is likely that other cellular functions will also be uncovered.

5
Summary and Perspectives

Histones are targeted by a variety of post-translational modifications that either increase or decrease the accessibility of DNA wrapped into chromatin. Histone monoubiquitylation is now emerging as an important modification for determining whether genes will be activated or silenced, and plays a role at several different steps in the transcription process. Like histone acetylation and methylation, histone ubiquitylation is regulated by evolutionarily conserved factors that act uniquely at each site targeted for this modification. Moreover, ubiquitin conjugation can be dynamic, and the sequential ubiquitylation and deubiquitylation of some histones during gene activation is a prerequisite for establishing optimal levels of transcription. Despite recent progress in the identification of the machinery that adds or removes ubiquitin on histones and the establishment of a role for histone ubiquitylation in transcriptional regulation, key questions remain to be answered. The primary question is how ubiquitin attachment influences transcription – does the ubiquitin moiety directly alter chromatin structure or does it provide an interaction surface to assemble or disassemble protein complexes that mediate different steps in transcription? A second related question concerns the mechanism by which monoubiquitylation of histone H2B regulates the trans-histone methylation of histone H3 and the function of this regulation in activated transcription. The past five years have seen many advances in our understanding of how this "ancient" histone modification is regulated and, given the rapid progress in this area, the next five years should yield answers to these important questions.

Acknowledgements The authors thank Shelley Berger, Danny Reinberg, and Yi Zhao for generously sharing unpublished data. Work from the authors' laboratory is supported by a grant from the NIH (GM40118) to M.A.O. and a Leukemia and Lymphoma Society Special Fellow award to C.-F.K.

References

Amar N, Messenguy F et al. (2000) ArgrII, a component of the Argr-Mcm1 complex involved in the control of arginine metabolism in Saccharomyces cerevisiae, is the sensor of arginine. Mol Cell Biol 20:2087–2097

Amerik AY, Li SJ et al. (2000) Analysis of the deubiquitinating enzymes of the yeast Saccharomyces cerevisiae. Biol Chem 381:981–992

Baarends WM, Hoogerbrugge JW et al. (1999) Histone ubiquitination and chromatin remodeling in mouse spermatogenesis. Dev Biol 207:322–333

Baarends WM, Wassenaar E et al. (2005) Silencing of unpaired chromatin and histone H2A ubiquitination in mammalian meiosis. Mol Cell Biol 25:1041–1053

Bailly V, Lauder S et al. (1997) Yeast DNA repair proteins Rad6 and Rad18 form a heterodimer that has ubiquitin conjugating, DNA binding, and ATP hydrolytic activities. J Biol Chem 272:23 360–23 365

Ballal NR, Kang YJ et al. (1975) Changes in nucleolar proteins and their phosphorylation patterns during liver regeneration. J Biol Chem 250:5921–5925

Bannister AJ, Schneider R et al. (2005) Spatial distribution of di- and tri-methyl lysine 36 of histone H3 at active genes. J Biol Chem 280:17 732–17 736

Barsoum J, Varshavsky A (1985) Preferential localization of variant nucleosomes near the 5′-end of the mouse dihydrofolate reductase gene. J Biol Chem 260:7688–7697

Bartel B, Wunning I et al. (1990) The recognition component of the n-end rule pathway. EMBO J 9:3179–3189

Bernstein BE, Humphrey EL et al. (2002) Methylation of histone H3 Lys 4 in coding regions of active genes. Proc Natl Acad Sci USA 99:8695–8700

Bhaumik SR, Green MR (2002) Differential requirement of SAGA components for recruitment of TATA-box-binding protein to promoters in vivo. Mol Cell Biol 22:7365–7371

Bohm L, Crane-Robinson C et al. (1980) Proteolytic digestion studies of chromatin core-histone structure. Identification of a limit peptide of histone H2A. Eur J Biochem 106:525–530

Bray S, Musisi H et al. (2005) Bre1 is required for notch signaling and histone modification. Dev Cell 8:279–286

Briggs SD, Xiao T et al. (2002) Gene silencing: trans-histone regulatory pathway in chromatin. Nature 418:498

Broomfield S, Hryciw T et al. (2001) DNA postreplication repair and mutagenesis in Saccharomyces cerevisiae. Mutat Res 486:167–184

Bryk M, Briggs SD et al. (2002) Evidence that Set1, a factor required for methylation of histone H3, regulates rDna silencing in S. cerevisiae by a Sir2-independent mechanism. Curr Biol 12:165–170

Buchberger A (2002) From Uba to Ubx: new words in the ubiquitin vocabulary. Trends Cell Biol 12:216–221

Cai SY, Babbitt RW et al. (1999) A mutant deubiquitinating enzyme (Ubp-m) associates with mitotic chromosomes and blocks cell division. Proc Natl Acad Sci USA 96:2828–2833

Cao R, Tsukuda YI, Zhang Y (2005) Role of Bmi-1 and Ring1A in H2A ubiquitylation and Hox gene silencing. Mol Cell 20:845–854

Cao R, Wang L et al. (2002) Role of histone H3 lysine 27 methylation in polycomb-group silencing. Science 298:1039–1043

Carvin CD, Kladde MP (2004) Effectors of lysine 4 methylation of histone H3 in Saccharomyces cerevisiae are negative regulators of Pho5 and Gal1-10. J Biol Chem 279:33 057–33 062

Chadwick BP, Willard HF (2003) Chromatin of the barr body: histone and non-histone proteins associated with or excluded from the inactive X chromosome. Hum Mol Genet 12:2167–2178

Chang M, French-Cornay D et al. (1999) A complex containing RNA polymerase II, Paf1p, Cdc73p, Hpr1p, and Ccr4p plays a role in protein kinase C signaling. Mol Cell Biol 19:1056–1067

Chen HY, Sun JM et al. (1998) Ubiquitination of histone H3 in elongating spermatids of rat testes. J Biol Chem 273:13 165–13 169

Cismowski MJ, Laff GM et al. (1995) Kin28 encodes a C-terminal domain kinase that controls mRNA transcription in Saccharomyces cerevisiae but lacks cyclin-dependent kinase-activating kinase (CAK) activity. Mol Cell Biol 15:2983–2992

Citterio E, Papait R et al. (2004) Np95 is a histone-binding protein endowed with ubiquitin ligase activity. Mol Cell Biol 24:2526–2535

Cohen HR, Royce-Tolland ME et al. (2005) Chromatin modifications on the inactive X chromosome. Prog Mol Subcell Biol 38:91–122

Costa PJ, Arndt KM (2000) Synthetic lethal interactions suggest a role for the Saccharomyces cerevisiae Rtf1 protein in transcription elongation. Genetics 156:535–547

Daniel JA, Torok MS et al. (2004) Deubiquitination of histone H2B by a yeast acetyltransferase complex regulates transcription. J Biol Chem 279:1867–1871

Davie JR, Lin R et al. (1991) Timing of the appearance of ubiquitinated histones in developing new macronuclei of Tetrahymena thermophila. Biochem Cell Biol 69:66–71

Davies N, Lindsey GG (1994) Histone H2B (and H2A) ubiquitination allows normal histone octamer and core particle reconstitution. Biochim Biophys Acta 1218:187–193

de Napoles M, Mermoud JE et al. (2004) Polycomb group proteins Ring1a/b link ubiquitylation of histone H2A to heritable gene silencing and X inactivation. Dev Cell 7:663–676

Dejardin J, Cavalli G (2005) Epigenetic inheritance of chromatin states mediated by Polycomb and Trithorax group proteins in Drosophila. Prog Mol Subcell Biol 38:31–63

Dong L, Xu CW (2004) Carbohydrates induce mono-ubiquitination of H2B in yeast. J Biol Chem 279:1577–1580

Dor Y, Raboy B et al. (1996) Role of the conserved carboxy-terminal alpha-helix of Rad6p in ubiquitination and DNA repair. Mol Microbiol 21:1197–1206

Dou Y, Milne TA et al. (2005) Physical association and coordinate function of the H3 K4 methyltransferase MLL1 and the H4 K16 acetyltransferase mof. Cell 121:873–885

Dover J, Schneider J et al. (2002) Methylation of histone H3 by compass requires ubiquitination of histone H2B by Rad6. J Biol Chem 277:28 368–28 371

Dudley AM, Rougeulle C et al. (1999) The Spt components of SAGA facilitate TBP binding to a promoter at a post-activator-binding step in vivo. Genes Dev 13:2940–2945

Emre NC, Berger SL (2004) Histone H2B ubiquitylation and deubiquitylation in genomic regulation. Cold Spring Harb Symp Quant Biol 69:289–299

Emre NC, Ingvarsdottir K et al. (2005) Maintenance of low histone ubiquitylation by Ubp10 correlates with telomere-proximal Sir2 association and gene silencing. Mol Cell 17:585–594

Ezhkova E, Tansey WP (2004) Proteasomal ATPases link ubiquitylation of histone H2B to methylation of histone H3. Mol Cell 13:435–442

Fang J, Chen T et al. (2004) Ring1b-mediated H2A ubiquitination associates with inactive X chromosomes and is involved in initiation of X inactivation. J Biol Chem 279:52 812–52 815

Fischle W, Wang Y et al. (2003) Histone and chromatin cross-talk. Curr Opin Cell Biol 15:172–183

Francis NJ, Kingston RE et al. (2004) Chromatin compaction by a Polycomb group protein complex. Science 306:1574–1577

Francis NJ, Saurin AJ et al. (2001) Reconstitution of a functional core Polycomb repressive complex. Mol Cell 8:545–556

Game JC, Williamson MS et al. (2005) X-ray survival characteristics and genetic analysis for nine Saccharomyces deletion mutants that show altered radiation sensitivity. Genetics 169:51–63

Gardner RG, Nelson ZW et al. (2005) Ubp10/Dot4p regulates the persistence of ubiquitinated histone H2B: distinct roles in telomeric silencing and general chromatin. Mol Cell Biol 25:6123–6139

Gerber M, Shilatifard A (2003) Transcriptional elongation by RNA polymerase II and histone methylation. J Biol Chem 278:26 303–26 306

Giannattasio M, Lazzaro F et al. (2005) The DNA damage checkpoint response requires histone H2B ubiquitination by Rad6-Bre1 and H3 methylation by Dot1. J Biol Chem 280:9879–9886

Goldknopf IL, Busch H (1977) Isopeptide linkage between nonhistone and histone 2A polypeptides of chromosomal conjugate-protein A24. Proc Natl Acad Sci USA 74:864–868

Goldknopf IL, French MF et al. (1978) A reciprocal relationship between contents of free ubiquitin and protein A24, its conjugate with histone 2A, in chromatin fractions obtained by the DNase II, Mg++ procedure. Biochem Biophys Res Commun 84:786–793

Goldknopf IL, Sudhakar S et al. (1980) Timing of ubiquitin synthesis and conjugation into protein A24 during the Hela cell cycle. Biochem Biophys Res Commun 95:1253–1260

Goldknopf IL, Taylor CW et al. (1975) Isolation and characterization of protein A24, a histone-like non-histone chromosomal protein. J Biol Chem 250:7182–7187

Greer SF, Zika E et al. (2003) Enhancement of CIITA transcriptional function by ubiquitin. Nat Immunol 4:1074–1082

Haglund K, Di Fiore PP et al. (2003) Distinct monoubiquitin signals in receptor endocytosis. Trends Biochem Sci 28:598–603

Hanson RD, Hess JL et al. (1999) Mammalian Trithorax and Polycomb-group homologues are antagonistic regulators of homeotic development. Proc Natl Acad Sci USA 96:14 372–14 377

Henry KW, Berger SL (2002) Trans-tail histone modifications: wedge or bridge? Nat Struct Biol 9:565–566

Henry KW, Wyce A et al. (2003) Transcriptional activation via sequential histone H2B ubiquitylation and deubiquitylation, mediated by SAGA-associated Ubp8. Genes Dev 17:2648–2663

Hernandez-Munoz I, Lund AH et al. (2005) Stable X chromosome inactivation involves the PRC1 polycomb complex and requires histone macroH2A1 and the Cullin3/Spop ubiquitin E3 ligase. Proc Natl Acad Sci USA 102:7635–7640

Hicke L (2001) Protein regulation by monoubiquitin. Nat Rev Mol Cell Biol 2:195–201

Hicke L, Dunn R (2003) Regulation of membrane protein transport by ubiquitin and ubiquitin-binding proteins. Annu Rev Cell Dev Biol 19:141–172

Hochstrasser M (1995) Ubiquitin, proteasomes, and the regulation of intracellular protein degradation. Curr Opin Cell Biol 7:215–223

Hoege C, Pfander B et al. (2002) Rad6-dependent DNA repair is linked to modification of PCNA by ubiquitin and sumo. Nature 419:135–141

Huang H, Kahana A et al. (1997) The ubiquitin-conjugating enzyme Rad6 (Ubc2) is required for silencing in Saccharomyces cerevisiae. Mol Cell Biol 17:6693–6699

Huyen Y, Zgheib O et al. (2004) Methylated lysine 79 of histone H3 targets 53bp1 to DNA double-strand breaks. Nature 432:406–411

Hwang WW, Venkatasubrahmanyam S et al. (2003) A conserved ring finger protein required for histone H2B monoubiquitination and cell size control. Mol Cell 11:261–266

Ingvarsdottir K, Krogan NJ et al. (2005) H2B ubiquitin protease Ubp8 and Sgf11 constitute a discrete functional module within the Saccharomyces cerevisiae SAGA complex. Mol Cell Biol 25:1162–1172

Jaenisch R, Beard C et al. (1998) Mammalian X chromosome inactivation. Novartis Found Symp 214:200–209; discussion 209–213, 228–232

Jason LJ, Finn RM et al. (2005) Histone H2A ubiquitination does not preclude histone H1 binding, but it facilitates its association with the nucleosome. J Biol Chem 280:4975–4982

Jason LJ, Moore SC et al. (2001) Magnesium-dependent association and folding of oligonucleosomes reconstituted with ubiquitinated H2A. J Biol Chem 276:14597–14601

Jason LJ, Moore SC et al. (2002) Histone ubiquitination: a tagging tail unfolds? Bioessays 24:166–174

Jentsch S, McGrath JP et al. (1987) The yeast DNA repair gene RAD6 encodes a ubiquitin-conjugating enzyme. Nature 329:131–134

Joazeiro CA, Weissman AM (2000) Ring finger proteins: mediators of ubiquitin ligase activity. Cell 102:549–552

Kang RS, Daniels CM et al. (2003) Solution structure of a CUE-ubiquitin complex reveals a conserved mode of ubiquitin binding. Cell 113:621–630

Kao CF, Hillyer C et al. (2004) Rad6 plays a role in transcriptional activation through ubiquitylation of histone H2B. Genes Dev 18:184–195

Kaplan CD, Holland MJ et al. (2005) Interaction between transcription elongation factors and mRNA 3′-end formation at the Saccharomyces cerevisiae GAL10-GAL7 locus. J Biol Chem 280:913–922

Koken M, Reynolds P et al. (1991) Dhr6, a Drosophila homolog of the yeast DNA-repair gene RAD6. Proc Natl Acad Sci USA 88:3832–3836

Koken MH, Hoogerbrugge JW et al. (1996) Expression of the ubiquitin-conjugating DNA repair enzymes hHR6a and b suggests a role in spermatogenesis and chromatin modification. Dev Biol 173:119–132

Koken MH, Reynolds P et al. (1991) Structural and functional conservation of two human homologs of the yeast DNA repair gene RAD6. Proc Natl Acad Sci USA 88:8865–8869

Kouzarides T (2002) Histone methylation in transcriptional control. Curr Opin Genet Dev 12:198–209

Krogan NJ, Dover J et al. (2003) The Paf1 complex is required for histone H3 methylation by COMPASS and Dot1p: linking transcriptional elongation to histone methylation. Mol Cell 11:721–729

Krogan NJ, Kim M et al. (2002) RNA polymerase II elongation factors of Saccharomyces cerevisiae: a targeted proteomics approach. Mol Cell Biol 22:6979–6992

Krogan NJ, Kim M et al. (2003) Methylation of histone H3 by Set2 in Saccharomyces cerevisiae is linked to transcriptional elongation by RNA polymerase II. Mol Cell Biol 23:4207–4218

Laribee RN, Krogan NJ et al. (2005) BUR kinase selectively regulates H3 K4 trimethylation and H2B ubiquitylation through recruitment of the PAF elongation complex. Curr Biol 15:1487–1493

Larschan E, Winston F (2001) The S. cerevisiae SAGA complex functions in vivo as a coactivator for transcriptional activation by Gal4. Genes Dev 15:1946–1956

Larschan E, Winston F (2005) The Saccharomyces cerevisiae Srb8-Srb11 complex functions with the SAGA complex during Gal4-activated transcription. Mol Cell Biol 25:114–123

Lee KK, Florens L et al. (2005) The deubiquitylation activity of Ubp8 is dependent upon Sgf11 and its association with the SAGA complex. Mol Cell Biol 25:1173–1182

Levinger L, Varshavsky A (1982) Selective arrangement of ubiquitinated and d1 protein-containing nucleosomes within the Drosophila genome. Cell 28:375–385

Lo WS, Henry KW et al. (2004) Histone modification patterns during gene activation. Methods Enzymol 377:130–153

Milne TA, Briggs SD et al. (2002) MLL targets set domain methyltransferase activity to Hox gene promoters. Mol Cell 10:1107–1117

Mimnaugh EG, Kayastha G et al. (2001) Caspase-dependent deubiquitination of monoubiquitinated nucleosomal histone H2A induced by diverse apoptogenic stimuli. Cell Death Differ 8:1182–1196

Minsky N, Oren M (2004) The ring domain of Mdm2 mediates histone ubiquitylation and transcriptional repression. Mol Cell 16:631–639

Montelone BA, Prakash S et al. (1981) Recombination and mutagenesis in rad6 mutants of Saccharomyces cerevisiae: evidence for multiple functions of the RAD6 gene. Mol Gen Genet 184:410–415

Moore SC, Jason L et al. (2002) The elusive structural role of ubiquitinated histones. Biochem Cell Biol 80:311–319

Morrison A, Miller EJ et al. (1988) Domain structure and functional analysis of the carboxyl-terminal polyacidic sequence of the Rad6 protein of Saccharomyces cerevisiae. Mol Cell Biol 8:1179–1185

Mueller RD, Yasuda H et al. (1985) Identification of ubiquitinated histones 2A and 2B in Physarum polycephalum. Disappearance of these proteins at metaphase and reappearance at anaphase. J Biol Chem 260:5147–5153

Mueller TD, Feigon J (2002) Solution structures of UBA domains reveal a conserved hydrophobic surface for protein–protein interactions. J Mol Biol 319:1243–1255

Muller J, Hart CM et al. (2002) Histone methyltransferase activity of a Drosophila Polycomb group repressor complex. Cell 111:197–208

Ng HH, Ciccone DN et al. (2003) Lysine-79 of histone H3 is hypomethylated at silenced loci in yeast and mammalian cells: a potential mechanism for position-effect variegation. Proc Natl Acad Sci USA 100:1820–1825

Ng HH, Dole S et al. (2003) The Rtf1 component of the PAF1 transcriptional elongation complex is required for ubiquitination of histone H2B. J Biol Chem 278:33 625–33 628

Ng HH, Feng Q et al. (2002) Lysine methylation within the globular domain of histone H3 by Dot1 is important for telomeric silencing and Sir protein association. Genes Dev 16:1518–1527

Ng HH, Robert F et al. (2003) Targeted recruitment of Set1 histone methylase by elongating Pol II provides a localized mark and memory of recent transcriptional activity. Mol Cell 11:709–719

Ng HH, Xu RM et al. (2002) Ubiquitination of histone H2B by Rad6 is required for efficient Dot1-mediated methylation of histone H3 lysine 79. J Biol Chem 277:34 655–34 657

Nickel BE, Allis CD et al. (1989) Ubiquitinated histone H2B is preferentially located in transcriptionally active chromatin. Biochemistry 28:958–963

Nickel BE, Roth SY et al. (1987) Changes in the histone H2A variant H2A.Z and polyubiquitinated histone species in developing trout testis. Biochemistry 26:4417–4421

Nislow C, Ray E et al. (1997) Set1, a yeast member of the Trithorax family, functions in transcriptional silencing and diverse cellular processes. Mol Biol Cell 8:2421–2436

O'Brien T, Tjian R (2000) Different functional domains of TAFII250 modulate expression of distinct subsets of mammalian genes. Proc Natl Acad Sci USA 97:2456–2461

Oh S, Zhang H et al. (2004) A mechanism related to the yeast transcriptional regulator PAF1c is required for expression of the Arabidopsis Flc/Maf Mads box gene family. Plant Cell 16:2940–2953

Orlandi I, Bettiga M et al. (2004) Transcriptional profiling of ubp10 null mutant reveals altered subtelomeric gene expression and insurgence of oxidative stress response. J Biol Chem 279:6414–6425

Orphanides G, Reinberg D (2000) Rna polymerase II elongation through chromatin. Nature 407:471–475

Pham AD, Sauer F (2000) Ubiquitin-activating/conjugating activity of TAFII250, a mediator of activation of gene expression in Drosophila. Science 289:2357–2360

Pickart CM (1997) Targeting of substrates to the 26s proteasome. Faseb J 11:1055–1066

Pickart CM (2001a) Mechanisms underlying ubiquitination. Annu Rev Biochem 70:503–533

Pickart CM (2001b) Ubiquitin enters the new millennium. Mol Cell 8:499–504

Pickart CM, Eddins MJ (2004) Ubiquitin: structures, functions, mechanisms. Biochim Biophys Acta 1695:55–72

Plath K, Fang J et al. (2003) Role of histone H3 lysine 27 methylation in X inactivation. Science 300:131–135

Pokholok DK, Hannett NM et al. (2002) Exchange of RNA polymerase II initiation and elongation factors during gene expression in vivo. Mol Cell 9:799–809

Pokholok DK, Harbison CT et al. (2005) Genome-wide map of nucleosome acetylation and methylation in yeast. Cell 122:517–527

Polo S, Sigismund S et al. (2002) A single motif responsible for ubiquitin recognition and monoubiquitination in endocytic proteins. Nature 416:451–455

Powell DW, Weaver CM et al. (2004) Cluster analysis of mass spectrometry data reveals a novel component of SAGA. Mol Cell Biol 24:7249–7259

Prag G, Misra S et al. (2003) Mechanism of ubiquitin recognition by the CUE domain of Vps9p. Cell 113:609–620

Prakash L (1981) Characterization of postreplication repair in Saccharomyces cerevisiae and effects of rad6, rad18, rev3 and rad52 mutations. Mol Gen Genet 184:471–478

Pray-Grant MG, Daniel JA et al. (2005) Chd1 chromodomain links histone H3 methylation with SAGA- and SLIK-dependent acetylation. Nature 433:434–438

Raasi S, Pickart CM (2005) Ubiquitin chain synthesis. Methods Mol Biol 301:47–55

Raboy B, Kulka RG (1994) Role of the C-terminus of Saccharomyces cerevisiae ubiquitin-conjugating enzyme (Rad6) in substrate and ubiquitin-protein-ligase (E3-r) interactions. Eur J Biochem 221:247–251

Rao B, Shibata Y, Strahl B, Lieb JD (2005) Dimethylation of histone H3 at lysine 30 demarcates regulatory and nonregulatory chromatin genome-wide. Mol Cell Biol 25:9447–9459

Reverol L, Chirinos M et al. (1997) Presence of an unusually high concentration of an ubiquitinated histone-like protein in Trypanosoma cruzi. J Cell Biochem 66:433–440

Reynolds P, Koken MH et al. (1990) The rhp6+ gene of Schizosaccharomyces pombe: a structural and functional homolog of the RAD6 gene from the distantly related yeast Saccharomyces cerevisiae. EMBO J 9:1423–1430

Ricci AR, Genereaux J et al. (2002) Components of the SAGA histone acetyltransferase complex are required for repressed transcription of ARG1 in rich medium. Mol Cell Biol 22:4033–4042

Robzyk K, Recht J et al. (2000) RAD6-dependent ubiquitination of histone H2B in yeast. Science 287:501–504

Roest HP, Baarends WM et al. (2004) The ubiquitin-conjugating DNA repair enzyme HR6A is a maternal factor essential for early embryonic development in mice. Mol Cell Biol 24:5485–5495

Roest HP, van Klaveren J et al. (1996) Inactivation of the HR6B ubiquitin-conjugating DNA repair enzyme in mice causes male sterility associated with chromatin modification. Cell 86:799–810

Ruppert S, Wang EH et al. (1993) Cloning and expression of human TAFII250: a TBP-associated factor implicated in cell-cycle regulation. Nature 362:175–179

Rusche LN, Kirchmaier AL et al. (2003) The establishment, inheritance, and function of silenced chromatin in Saccharomyces cerevisiae. Annu Rev Biochem 72:481–516

San-Segundo PA, Roeder GS (2000) Role for the silencing protein Dot1 in meiotic checkpoint control. Mol Biol Cell 11:3601–3615

Santos-Rosa H, Schneider R et al. (2002) Active genes are tri-methylated at K4 of histone H3. Nature 419:407–411

Schneider J, Wood A et al. (2005) Molecular regulation of histone H3 trimethylation by COMPASS and the regulation of gene expression. Mol Cell 19:849–856

Seale RL (1981) Rapid turnover of the histone–ubiquitin conjugate, protein A24. Nucleic Acids Res 9:3151–3158

Shahbazian MD, Zhang K et al. (2005) Histone H2B ubiquitylation controls processive methylation but not monomethylation by Dot1 and Set1. Mol Cell 19:271–277

Shi X, Chang M et al. (1997) Cdc73p and Paf1p are found in a novel RNA polymerase II-containing complex distinct from the Srbp-containing holoenzyme. Mol Cell Biol 17:1160–1169

Shi X, Finkelstein A et al. (1996) Paf1p, an RNA polymerase II-associated factor in Saccharomyces cerevisiae, may have both positive and negative roles in transcription. Mol Cell Biol 16:669–676

Shih SC, Prag G et al. (2003) A ubiquitin-binding motif required for intramolecular monoubiquitylation, the CUE domain. EMBO J 22:1273–1281

Sigismund S, Polo S et al. (2004) Signaling through monoubiquitination. Curr Top Microbiol Immunol 286:149–185

Simic R, Lindstrom DL et al. (2003) Chromatin remodeling protein Chd1 interacts with transcription elongation factors and localizes to transcribed genes. EMBO J 22:1846–1856

Singh J, Goel V et al. (1998) A novel function of the DNA repair gene rhp6 in mating-type silencing by chromatin remodeling in fission yeast. Mol Cell Biol 18:5511–5522

Smith KP, Byron M et al. (2004) Ubiquitinated proteins including uH2A on the human and mouse inactive X chromosome: enrichment in gene rich bands. Chromosoma 113:324–335

Sollier J, Lin W et al. (2004) SET1 is required for meiotic s-phase onset, double-strand break formation and middle gene expression. EMBO J 23:1957–1967

Squazzo SL, Costa PJ et al. (2002) The PAF1 complex physically and functionally associates with transcription elongation factors in vivo. EMBO J 21:1764–1774

Stelter P, Ulrich HD (2003) Control of spontaneous and damage-induced mutagenesis by SUMO and ubiquitin conjugation. Nature 425:188–191

Sterner DE, Grant PA et al. (1999) Functional organization of the yeast SAGA complex: distinct components involved in structural integrity, nucleosome acetylation, and TATA–binding protein interaction. Mol Cell Biol 19:86–98

Sun ZW, Allis CD (2002) Ubiquitination of histone H2B regulates H3 methylation and gene silencing in yeast. Nature 418:104–108

Sung P, Prakash S et al. (1988) The Rad6 protein of Saccharomyces cerevisiae polyubiquitinates histones, and its acidic domain mediates this activity. Genes Dev 2:1476–1485

Sung P, Prakash S et al. (1990) Mutation of cysteine-88 in the Saccharomyces cerevisiae RAD6 protein abolishes its ubiquitin-conjugating activity and its various biological functions. Proc Natl Acad Sci USA 87:2695–2699

Swerdlow PS, Schuster T et al. (1990) A conserved sequence in histone H2A which is a ubiquitination site in higher eucaryotes is not required for growth in Saccharomyces cerevisiae. Mol Cell Biol 10:4905–4911

Tasaki T, Mulder LC et al. (2005) A family of mammalian E3 ubiquitin ligases that contain the UBR box motif and recognize N-degrons. Mol Cell Biol 25:7120–7136

Thorne AW, Sautiere P et al. (1987) The structure of ubiquitinated histone H2B. EMBO J 6:1005–1010

Turner SD, Ricci AR et al. (2002) The E2 ubiquitin conjugase rad6 is required for the ArgR/Mcm1 repression of ARG1 transcription. Mol Cell Biol 22:4011–4019

Ulrich HD (2004) How to activate a damage-tolerant polymerase: consequences of PCNA modifications by ubiquitin and SUMO. Cell Cycle 3:15–18

Ulrich HD, Jentsch S (2000) Two ring finger proteins mediate cooperation between ubiquitin-conjugating enzymes in DNA repair. EMBO J 19:3388–3397

van der Knaap JA, Kumar BR et al. (2005) Gmp synthetase stimulates histone H2B deubiquitylation by the epigenetic silencer Usp7. Mol Cell 17:695–707

van Leeuwen F, Gafken PR et al. (2002) Dot1p modulates silencing in yeast by methylation of the nucleosome core. Cell 109:745–756

van Leeuwen F, Gottschling DE (2002) Genome-wide histone modifications: gaining specificity by preventing promiscuity. Curr Opin Cell Biol 14:756–762

Vassilev AP, Rasmussen HH et al. (1995) The levels of ubiquitinated histone H2A are highly upregulated in transformed human cells: partial colocalization of uH2A clusters and PCNA/cyclin foci in a fraction of cells in S-phase. J Cell Sci 108 (Part 3):1205–1215

Wang H, Wang L et al. (2004) Role of histone H2A ubiquitination in Polycomb silencing. Nature 431:873–878

Wang L, Brown JL et al. (2004) Hierarchical recruitment of Polycomb group silencing complexes. Mol Cell 14:637–646

Watkins JF, Sung P et al. (1993) The extremely conserved amino terminus of RAD6 ubiquitin-conjugating enzyme is essential for amino-end rule-dependent protein degradation. Genes Dev 7:250–261

West MH, Bonner WM (1980) Histone 2B can be modified by the attachment of ubiquitin. Nucleic Acids Res 8:4671–4680

Wilkinson KD (1997) Regulation of ubiquitin-dependent processes by deubiquitinating enzymes. Faseb J 11:1245–1256

Wing SS, Jain P (1995) Molecular cloning, expression and characterization of a ubiquitin conjugation enzyme (E2(17)kb) highly expressed in rat testis. Biochem J 305 (Part 1):125–132

Wood A, Krogan NJ et al. (2003) Bre1, an E3 ubiquitin ligase required for recruitment and substrate selection of Rad6 at a promoter. Mol Cell 11:267–274

Wood A, Schneider J et al. (2003) The PAF1 complex is essential for histone monoubiquitination by the Rad6/Bre1 complex, which signals for histone methylation by COMPASS and Dot1p. J Biol Chem 278:34739–34742

Wood A, Schneider J et al. (2005) The Bur1/Bur2 complex is required for H2B monoubiquitination by Rad6/Bre and histone methylation by COMPASS. Mol Cell 20:589–599

Wu RS, Kohn KW et al. (1981) Metabolism of ubiquitinated histones. J Biol Chem 256:5916–5920

Wysocka J, Swigut T et al. (2005) Wdr5 associates with histone H3 methylated at K4 and is essential for H3 K4 methylation and vertebrate development. Cell 121:859–872

Wysocki R, Javaheri A et al. (2005) Role of Dot1-dependent histone H3 methylation in G1 and S phase DNA damage checkpoint functions of Rad9. Mol Cell Biol 25:8430–8443

Xiao T, Hall H et al. (2003) Phosphorylation of RNA polymerase III CTD regulates H3 methylation in yeast. Genes Dev 17:654–663

Xiao T, Kao CF et al. (2005) Histone H2B ubiquitylation is associated with elongating RNA polymerase II. Mol Cell Biol 25:637–651

Xie Y, Varshavsky A (1999) The E2-E3 interaction in the N-end rule pathway: the RING-H2 finger of E3 is required for the synthesis of multiubiquitin chain. EMBO J 18:6832–6844

Yamashita K, Shinohara M et al. (2004) Rad6-Bre1-mediated histone H2B ubiquitylation modulates the formation of double-strand breaks during meiosis. Proc Natl Acad Sci USA 101:11 380–11 385

Yao S, Neiman A et al. (2000) BUR1 and BUR2 encode a divergent cyclin-dependent kinase–cyclin complex important for transcription in vivo. Mol Cell Biol 20:7080–7087

Ye J, Ai X et al. (2005) Histone H4 lysine 91 acetylation, a core domain modification associated with chromatin assembly. Mol Cell 18:123–130

Zhang Y (2003) Transcriptional regulation by histone ubiquitination and deubiquitination. Genes Dev 17:2733–2740

Zhang Y, Cao R et al. (2004) Mechanism of Polycomb group gene silencing. Cold Spring Harb Symp Quant Biol 69:309–317

Zhu B, Mandal SS et al. (2005) The human PAF complex coordinates transcription with events downstream of RNA synthesis. Genes Dev 19:1668–1673

Zhu B, Zheng Y et al. (2005) Monoubiquitination of human histone H2B: the factors involved and their roles in Hox gene regulation. Mol Cell 20:601–611

Results Probl Cell Differ (41)
L. Brehon: Chromatin Dynamics in Cellular Function
DOI 10.1007/009/Published online: 15 February 2006
© Springer-Verlag Berlin Heidelberg 2006

Histone Dynamics During Transcription: Exchange of H2A/H2B Dimers and H3/H4 Tetramers During Pol II Elongation

Christophe Thiriet[1,2] · Jeffrey J. Hayes[1] (✉)

[1]Department of Biochemistry and Biophysics, University of Rochester Medical Center,
 Box 712, Rochester, NY 14625, USA
 jjhs@mail.rochester.edu

[2]*Present address:*
 Institut Andre Lwoff, CNRS UPR-1983, 94801 Villejuif cedex, France

Abstract Chromatin within eukaryotic cell nuclei accommodates many complex activities that require at least partial disassembly and reassembly of nucleosomes. This disassembly/reassembly is thought to be somewhat localized when associated with processes such as site-specific DNA repair but likely occurs over extended regions during processive processes such as DNA replication or transcription. Here we review data addressing the effect of transcription elongation on nucleosome disassembly/reassembly, specifically focusing on the issue of transcription-dependent exchange of H2A/H2B dimers and H3/H4 tetramers. We suggest a model whereby passage of a polymerase through a nucleosome induces displacement of H2A/H2B dimers with a much higher probability than displacement of H3/H4 tetramers such that the extent of tetramer replacement is relatively low and proportional to polymerase density on any particular gene.

1
A Brief History of Chromatin and Transcription

Understanding the mechanisms by which RNA polymerases access DNA in the refractory environment of chromatin has been the focus of many laboratories for decades. It was recognized in the early 1960s that association of histones with DNA severely restricts the utilization of the molecule as a template for biological processes such as transcription (Huang and Bonner 1962; Silverman and Mirsky 1973). Later experiments indicated that both transcription initiation and elongation can be blocked by the presence of nucleosomes on the DNA (Knezetic and Luse 1986; Izban and Luse 1991; Wolffe and Kurumizaka 1998) suggesting that efficient transcription in vivo might require significant alteration of native chromatin structure. Indeed, mapping of heat shock loci using DNAase I and MNase revealed that transcription activity is correlated with an opening of chromatin, evident by the loss of the normal nucleosomal pattern and the appearance of sites hypersensitive to cleavage by nucleases (Wu et al. 1979; Wu 1980). These alterations were typically found localized near the 5′ ends of the genes (Wu 1980) suggesting that opening is primarily associated with enhancer and promoter binding by trans-acting factors and transcription initiation rather than polymerase elongation.

Alteration of chromatin structure may be brought about by several mechanisms including effects of posttranslational modifications of the histone proteins, inclusion of histone variants within the chromatin and the activity of ATP-dependent chromatin remodeling factors. Evidence for the role of histone posttranslational modifications in gene transcription was first provided by Allfrey and co-workers who showed that histone proteins were acetylated and that acetylation reduces the ability of histones to inhibit transcription by RNA polymerase activities (Allfrey et al. 1964). Over 30 years later, Allis and co-workers provided a direct mechanistic link between histone acetylation and gene transcription by demonstrating that the transcription co-activator Gcn5p is a histone acetyltransferase (Brownell et al. 1996). Since then much work has firmly established a critical role for histone posttranslational modifications, including acetylation, in both transcription initiation and elongation (Grunstein 1997; Strahl and Allis 2000; Zhang and Reinberg 2001; Peterson and Laniel 2004).

Current models suggest that histone posttranslational modifications elicit both direct and indirect effects on chromatin structure. For example, acetylation reduces the ability of nucleosome arrays to undergo salt-dependent folding into secondary and tertiary chromatin structures (Annunziato et al. 1988; Woodcock and Dimitrov 2001; Hansen 2002). However the molecular mechanisms by which acetylation directly modulates chromatin folding remain unclear (Zheng and Hayes 2003). In addition, posttranslational modifications provide recognition signals that direct the binding of critical trans-acting factors. Individual and multiple acetylations are recognized by bromodomain-containing factors, resulting in the association of other chromatin modifying activities (de la Cruz et al. 2005). These activities include ATP-dependent chromatin remodeling complexes able to alter the conformation and positions nucleosomes, allowing freer access to the underlying DNA (Boyer et al. 2000; Peterson 2000).

In some cases chromatin remodeling, in conjunction with histone chaperones or assembly factors, can result in the complete removal of all or part of the core histone octamer from the DNA within promoter regions (Boeger et al. 2003; Reinke and Horz 2003; Adkins et al. 2004). Similar mechanisms likely account for the nuclease sensitivity of promoter regions observed in earlier experiments (Wu 1980), and indicate that in general promoter structure is drastically altered to render the chromatin structure permissive for the binding of transcription factors and co-activators via a combination of posttranslational modifications of histones, chromatin remodeling, and the binding of specific ancillary factors (Cosma 2002).

2
RNA Polymerase Activity Induces Histone Exchange with Free Pools

In addition to biochemical studies of promoter architecture at active loci, several other lines of investigation have suggested that alterations in chromatin structure occur in association with transcription. During S-phase, nascent H3/H4 tetramers are found on nascent DNA, as expected, but are associated with both old and new H2A/H2B dimers, suggesting that dimers are incorporated throughout the S-phase nucleus via both replication and transcription-related mechanisms (Jackson et al. 1981). Early evidence for transcription-dependent incorporation of nascent histones into chromatin in vivo was provided by pulse chase experiments performed at specific cell cycle stages in mammalian cell cultures (Jackson and Chalkley 1985; Louters and Chalkley 1985). These experiments showed that histones are not exclusively synthesized in S-phase, but about 5% of core histones are synthesized in G1-phase, raising the possibility that histone synthesis occurs according to both a replication dependent and a replication independent regime. Interestingly, these investigators found that the patterns of incorporation of histones into chromatin also depended on cell cycle stage and histone type. By coupling pulse-chase labeling of cells, density gradients and electrophoretic analyses, Jackson and Chalkley showed that in G1 phase more nascent H2A/H2B dimers synthesized in a replication-independent fashion were associated with DNA than nascent H3/H4 (Jackson and Chalkley 1985). Although direct evidence linking the replication independent assembly with transcription was not provided, it seems reasonable to hypothesize that transcription facilitated incorporation of nascent H2A/H2B dimers into chromatin, while this histone exchange takes place much less frequently for H3/H4.

Further support for transcription-induced incorporation of nascent H2A/H2B dimers into chromatin was provided by Annunziato and colleagues who performed early chromatin immunoprecipitation experiments following treatment of cells with hydroxyurea (a replication inhibitor) in conjunction with pulse labeling (Perry et al. 1993). This analysis revealed that a substantial fraction of nascent H2A/H2B associated with mature chromatin regions containing highly acetylated H4, suggesting that dimers are preferentially assembled into transcriptionally active regions of chromatin. Again, although no direct evidence that H2A/H2B incorporation was directly associated with RNA polymerase activity was available, at the very least these data indicate that the disruption/reformation of nucleosomes in vivo occurs at sites where chromatin is heavily acetylated.

Evidence for transcription-dependent histone exchange has also been obtained with chromatin fractionated based on solubility in 150 mM NaCl from immature erythrocytes (Hendzel and Davie 1990). In concurrence with the above-mentioned studies, these workers found exchange of nascent H2A/H2B occurred on regions that fractionated with actively transcribed genes. In addition, they found that traces of nascent H3/H4 (specifically H3.3) were associated with this same fraction suggesting that the non-replication related H3 variant H3.3 is incorporated into chromatin in connection with transcription (see Sect. 4, below). However, in this study a similar level of incorporation was found in chromatin from cells treated with the transcription inhibitor actinomycin D, suggesting that the observed exchange is due to an inherent reduced stability of nucleosomes within active/competent regions of chromatin and not directly related to the active process of transcription. Alternatively, actinomycin D itself may induce some histone exchange via direct alterations of chromatin structure that might mask effects due to transcription inhibition (Kimura and Cook 2001).

In sum, all these analyses clearly demonstrate that chromatin is a dynamic complex, with assembly and disassembly occurring throughout the cell cycle in a manner dependent upon transcription and perhaps other active processes. In addition, these results indicate that the free pool of histones needs to be constantly replenished with nascent proteins even during G1 and G2 phase, thus processes causing histone displacement and replacement result in incorporation of nascent histones into the chromatin. Moreover, histone displacement must occur in such a way that the displaced histones are not available for immediate reassembly onto the same positions in the chromatin, perhaps due to dilution into a free pool that also contains the nascent proteins. Clearly, the experiments examining chromatin assembly during S-phase mentioned above suggest that a significant fraction of the H2A/H2B dimers displaced during transcription are "recycled" as part of the free histone pool. Alternatively, it is possible that the reassembly of transcription-displaced histones occurs at a much slower rate than the deposition of nascent proteins delivered from the cytoplasm. Nevertheless, regardless of the actual mechanisms involved, these experiments suggest that replication-independent turnover of H2A/H2B dimers occurs with a much higher frequency than turnover of H3/H4 tetramers.

3

Histone Exchange May be Due to RNA Pol II Elongation Through Nucleosomes

In contrast to the mechanisms by which gene promoters are made accessible for binding of trans-acting factors and initiation of transcription, the mechanisms by which RNA polymerases elongates through chromatinized templates are less well understood. Examination of templates transcribed in vitro indicated that while nucleosomes typically blocked or severely reduced elongation by RNA polymerase II, smaller bacteriophage polymerases were able to transcribe through nucleosomes, albeit at reduced rates (see Chang and Luse 1997 and references therein). Although these studies indicated that additional factors were required for efficient pol II elongation in vivo, studies of the more experimentally tractable bacteriophage polymerases revealed the first clues of how transcription could occur on nucleosome templates. In a pioneering experiment, Clark and Felsenfeld inserted a DNA fragment bearing a single nucleosome into a plasmid downstream of a SP6 promoter (Clark and Felsenfeld 1992). They observed that upon transcription of the region, the nucleosome was transferred to other locations within the plasmid, with a strong bias for the region immediately upstream of the promoter.

To more directly address the mechanism by which RNA polymerases pass a nucleosomal barrier, Studitsky, Felsenfeld and colleagues reconstituted a single positioned nucleosome on a specific transcribeable DNA fragment. They found that both RNA polymerase III and SP6 polymerase were able to transcribe through the nucleosomal region, and that the histone octamer was transferred to a position upstream from its original location (Studitsky et al. 1994, 2004; Felsenfeld et al. 2000). Moreover, close inspection of RNA polymerase pausing during transit through the nucleosome suggested that the nucleosomal DNA was slowly unwrapped from the histone surface until about the nucleosome dyad, then the remainder of the template rapidly transcribed (Studitsky et al. 1995). Coupled with evidence suggesting that the octamer did not completely dissociate from the DNA template during the process, these authors proposed a model whereby polymerase invasion of the nucleosome results in the formation of a bulge or loop containing the polymerase followed by directed transfer of the histone octamer to a location behind the advancing polymerase (Felsenfeld et al. 2000; Studitsky et al. 2004). It is important to note that both bacterial polymerase and pol III were observed to transfer the entire histone octamer out of the path of the polymerase to a position behind its original location, suggesting that the octamer transfer does not involve a free intermediate (Studitsky et al. 2004).

Obviously, these experiments did not explain why transcription was observed to induce the mobilization of H2A/H2B in vivo. However, the smaller sizes of SP6 and Pol III polymerases, raised the possibility that the situation might be different with pol II. Indeed, this turns out to be the case. Recent studies by Studitsky and collaborators investigated the mechanism of transcription by RNA polymerase II through nucleosomes in the same manner (Kireeva et al. 2002, 2005; Belotserkovskaya et al. 2003). A key to these studies was an elegant methodology whereby a template containing a stalled RNA pol II is ligated to a nucleosome, followed by purification of active nucleoso mal complexes (Kireeva et al. 2002). Importantly, unlike RNA pol III and SP6 polymerase, transcription by RNA pol II in the reconstituted system showed that one histone dimer of H2A/H2B was displaced from the octamer upon polymerase passage. Moreover, the complex FACT, previously shown to stimulate pol II elongation on nucleosome templates (Orphanides et al. 1998), was found to stimulate dimer displacement in this experiment (Belotserkovskaya et al. 2003). Furthermore, Spt16, a subunit of *Drosophila* FACT, co-localizes with elongation factors and RNA polymerase in polytene chromosomes and heat-shock gene induction in polytene chromosomes showed that FACT is recruited to sites of active transcription (Saunders et al. 2003). Interestingly, there is no evidence for FACT recruitment to genes transcribed by RNA polymerase III suggesting that the mechanism by which transcription elongation occurs on these genes may be different than pol II genes, in agreement with in vitro studies (Studitsky et al. 1997).

Support for the idea that transcription-dependent H2A/H2B dimer exchange in vivo is primarily due to transcription elongation rather than promoter remodeling or events associated with initiation was recently presented (Thiriet and Hayes 2005). These authors exploited the natural synchrony of millions of nuclei within the unicellular macroplasmodium of the slime mold *Physarum polycelphalum* and the ability of this organism to internalize exogenous proteins into its cellular metabolism to examine histone incorporation outside of S-phase. By introducing epitope-tagged exogenous histones directly into the cell, they used ChIP to show H2A/H2B dimer exchange was readily observed on transcribed genes but was much less prevalent on silent loci. Moreover, exchange was preferentially confined to the structural gene within transcribed loci suggesting that active elongation causes exchange of H2A/H2B dimers with free pools rather than increased levels of passive exchange within transcriptionally active, highly acetylated open chromatin domains (Thiriet and Hayes 2005). Interestingly, significantly less exchange of histones H3/H4 was found on the active loci (see below). These results support the model of Studitsky and colleagues and also suggest that such "dimer-displaced" nucleosomes may be related to the partial nucleosome structures observed after pol II transcription of chromatin templates in vivo in yeast cells (Sathyanarayana et al. 1999).

4
Exchange of H3/H4 Tetramers During Transcription

As mentioned above, early experiments identified sets of both H2A/H2B and H3/H4 histones that were synthesized in a replication-independent fashion (Jackson and Chalkley 1985). However, while replication-independent incorporation of H2A/H2B into chromatin has been well established, incorporation of H3/H4 outside of S-phase has been much harder to detect (Jackson and Chalkley 1985). For example, recent experiments examining mobility of core histone-GFP fusions found that while a fraction ($\sim$ 3%) of H2A/H2B within mammalian cell nuclei appears to be very mobile in a transcription-dependent fashion, bulk H3-GFP and H4-GFP did not exhibit similar mobility (Kimura and Cook 2001). However, as mentioned above, Hendzel and Davie detected nascent H3/H4, specifically tetramers containing the histone variant H3.3, in active chromatin fractionated based on solubility in NaCl solutions

(Hendzel and Davie 1990). Thus the apparent lack of detection of replication-independent incorporation of H3/H4 may be in part due to the fact that the majority of this exchange involves the specialized H3 variant H3.3 (see below). However, it is important to note that the experiments of Kimura and Cook (2001) examined the mobility of both H3-GFP and H4-GFP, suggesting that quantitative replacement with nascent H3.3/H4-GFP should have been detected in these cells.

H3.3 is a non-allelic H3 variant typically constitutively synthesized in low amounts and thus has been referred to as a replacement variant (Thatcher et al. 1994). Despite being widely distributed throughout eukaryotes, H3.3s appear to have evolved independently (Thatcher et al. 1994) but appear to play a common role in replication-independent chromatin assembly (Yu and Gorovsky 1997; Ahmad and Henikoff 2002; Wirbelauer et al. 2005). Henikoff and colleagues have shown that H3.3 in *Drosophila* can be assembled into chromatin in both replication-coupled and replication-independent processes, while major H3s are excluded from deposition in the latter (Ahmad and Henikoff 2002). Thus some H3/H4 tetramers containing S-phase synthesized H3.1 and H3.2 are gradually replaced with tetramers containing H3.3 during G1 and G2 phases. Furthermore, they hypothesize that the H3.3 tetramers "mark" active chromatin in some manner, perhaps to facilitate subsequent rounds of transcription. This work clearly showed that polymerase I induces such an exchange mechanism since a majority of accumulation of the H3.3-GFP fusion was within the rDNA locus (Ahmad and Henikoff 2002). More recent results indicate that H3.3 also is deposited onto active pol II transcription units (Mito et al. 2005) and is enriched in modifications associated with active loci (McKittrick et al. 2004; Hake et al. 2005).

5
H2A/H2B vs H3/H4 Exchange

As mentioned above, methods examining global histone exchange detect little or no exchange of bulk H3/H4 compared to H2A/H2B, possibly due to the fact that while H3.3/H4 tetramers are assembled into chromatin in a transcription-dependent manner, methods that monitor redistribution of the major-type H3s do not detect turnover due to H3.3 incorporation. In addition, it is likely that the level of transcription-associated H3.3 incorporation is far less than that which occurs for H2A/H2B. Indeed, a direct comparison of exchange of H2A/H2B and H3/H4 (including H3.3) on active loci in vivo in the slime mold *Physarum* found that exchange of H2A/H2B occurs > 20 times more frequently than exchange of H3/H4 (Thiriet and Hayes 2005). Thus, while each passage of the polymerase may induce release of a H2A/H2B dimer with high probability, replacement of tetramers may occur with significantly less frequency and possibly via a distinct mechanism. Indeed, recent work from Henikoff's lab shows that H3.3 appears to be distributed far upstream and downstream of active loci (Mito et al. 2005) while H2A/H2B exchange appears to be limited to the transcribed unit (Thiriet and Hayes 2005). While this wide-spread H3.3 was attributed to intergenic transcription, it is also possible that H3.3 incorporation occurs via a mechanism linked to locus control (Hendzel and Davie 1990) (see above).

Several other pieces of evidence support the idea that H3.3/H4 exchange may be induced by transcription but with a reduced probability compared to H2A/H2B exchange. For example, real time analyses of human cell cultures showed that H3.3 deposition into a site highly transcribed by RNA pol II is delayed compared to the accumulation of transcripts (Janicki et al. 2004). Also, in contrast to histone exchange at loci transcribed by RNA pol II, it was found that in nucleolar DNA H3/H4 were rapidly displaced in regions transcribed by RNA pol I in vivo (Thiriet and Hayes 2005). These data suggest that displacement of the H3/H4 tetramer may depend on the rate of transcription and/or polymerase density, an idea supported by a recent analysis of ectopically expressed H3.3 in cultured *Drosophila* Kc cells (Wirbelauer et al. 2005). These authors also showed that while histone posttranslational modifications associated with transcription are biased toward the 5′ end of genes, H3.3 appears to be evenly distributed across active genes in proportion to the concentration of elongating polymerase (Wirbelauer et al. 2005). In addition to the rate of incorporation, the extent of replacement of H2A/H2B versus H3/H4 with proteins from the free pool may be quite different. Experiments in *Physarum* suggest that nearly 100% of H2A/H2B dimers are exchanged within 3 hrs while ≤ 5% of tetramers exchanged during the same time period on moderately transcribed "housekeeping" genes. In Kc cells, upon cessation of induced transcription of a highly transcribed reporter, Wirbelauer et al. found that the ability to ChIP major H3 is reduced only about twofold while that of H3.3 is increased by twofold, suggesting that only partial replacement of H3.1 or H3.2 with H3.3 on the transcribed unit had occurred. However, analysis of the abundance of H3.3 in *Drosophila* Kc cells indicated that this variant comprises 25% of all H3 within the cell, enough to package all active loci (McKittrick et al. 2004). Thus more data are needed regarding the level of H3.3 and extent of replacement of major type H3s on actively transcribed genes in different tissues and organisms.

A significant factor yet to be considered in most in vitro model systems examining the passage of RNA polymerases through nucleosomes is contribution of torsional stress within the DNA. Unwinding of the double-stranded DNA by RNA polymerase coupled with the fact that a large polymerase complex trailing a nascent RNA transcript does not freely diffuse around the double helical template causes positive superhelical stress to accumulate in front of the advancing polymerase and negative stress behind (Giaever and Wang 1988; Tsao et al. 1989). Recent work from Jackson's laboratory indicates that such superhelical stress can have profound effects on the retention and displacement of histones from the DNA. For example, the data suggest that positive stress alone may cause the ejection of an H2A/H2B dimer from a nucleosome but also may induce a positively coiled DNA structure that preferentially retains H3/H4 tetramers on the transcribed DNA template (Levchenko et al. 2005). In this scenario, the tetramer would "flip-handedness" in such a manner that it would constrain positive rather than negative supercoils in the DNA (Alilat et al. 1999). In addition, evidence suggests that histone chaperones can significantly modulate the effect of torsional stress on histone displacement (Levchenko et al. 2005). Clearly, mechanistic studies in the future will have to recapitulate the torsionally constrained environment in which transcription must occur.

6
Perspectives

We propose a model whereby transcription of a polymerase through a nucleosome leads to displacement of an H2A/H2B dimer (see Fig. 1), consistent with recent in vivo and in vitro experiments (Kireeva et al. 2002, 2005; Thiriet and Hayes 2005). Although in vitro experiments employed only a single pass of polymerase and resulted in loss of only a single H2A/H2B dimer in a fraction of the nucleosomes (Kireeva et al. 2002; Belotserkovskaya et al. 2003), in vivo experiments suggest that upon multiple polymerase passages eventually both dimers are displaced and replaced by nascent proteins from the free pool (Thiriet and Hayes 2005). It is worth noting that at this point it is unclear whether within a single nucleosome the first or second dimers encountered by the polymerase are displaced with similar probabilities. Nevertheless, our model predicts that at low rates of transcription, dimer exchange occurs with a high probability while displacement of entire nucleosomes and transcription-dependent incorporation of H3.3/H4 tetramers occurs to a much less frequent but detectable extent (Mito et al.; Wirbelauer et al.). At high rates of transcription, displacement of whole nucleosomes occurs more frequently, perhaps because RNA polymerase density accumulates beyond a threshold that no longer allows nucleosome transfer around the preceding polymerase. This leads to much more frequent transcription-dependent incorporation of H3.3/H4.

In vivo experiments, indicate that much more transcription-induced H3.3 incorporation occurs on pol I than pol II genes. However, to date no analyses have been carried out with RNA pol I in vitro. The primary differences in the mechanism between pol II and pol I might be in the rates of transcription. Thus future experiments in vitro should explore whether displacement of entire nucleosomes occurs when RNA polymerase densities exceed threshold levels that would sterically restrict nucleosome reformation. Indeed recent ChIP experiments examining the distribution of nucleosomes throughout the

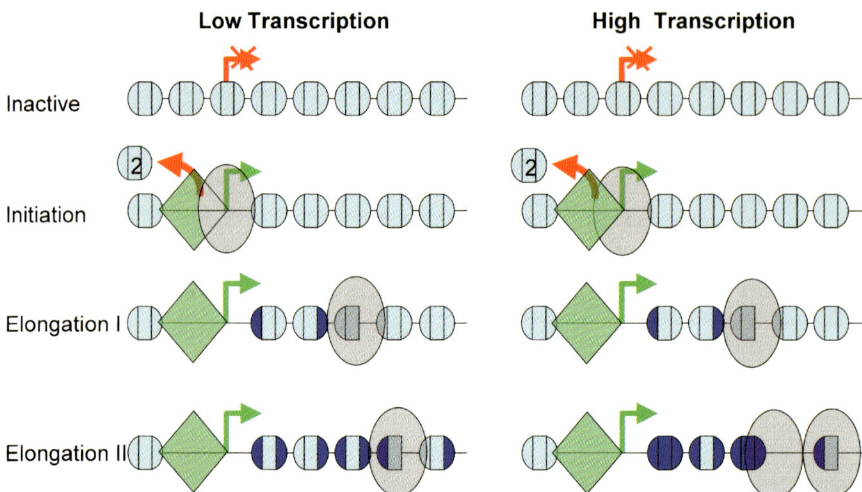

Fig. 1 A model for polymerase II-induced exchange of H2A/H2B dimers and H3/H4 tetramers is presented for genes undergoing moderate and high levels of transcription. H2A/H2B dimers and H3/H4 tetramers are shown as *semi-circles* and *rectangles*, respectively. Chromatin proteins replaced as a result of transcription are *colored blue*. Transcription factors binding to the remodeled promoter are depicted collectively as a *green diamond* and pol II is shown as a *grey oval*. The start site of transcription is indicated (*arrow*). Note that two nucleosomes are displaced as a result of pre-initiation complex formation. The model depicts that under conditions of low transcription, elongation induces mainly partial nucleosome disruption resulting in displacement of H2A/H2B dimers, while displacement of entire nucleosomes and incorporation of nascent H3.3/H4 tetramers occurs with a much lower probability. High levels of transcription lead to higher polymerase densities and more frequent displacement of entire nucleosomes and more frequent incorporation of H3.3/H4 tetramers. This may be due to interference of hexamer transfer around a polymerase due to a closely following second polymerase

yeast genome and at specific loci have revealed that transcribed sequences of very active regions are deficient of core histones (Lee et al. 2004; Schwabish and Struhl 2004) suggesting that high densities of transcribing polymerases can force detectable displacement of whole nucleosomes and, in higher organisms, eventual replacement with H3.3/H4 tetramers (note that in yeast all H3 is comprised of H3.3). Regardless of the mechanisms involved, the in vivo and in vitro data suggest that dimer and tetramer displacement/exchange may occur via a distinct mechanisms. Moreover, limited tetramer exchange would provide for greater conservation of epigenetic marks on the H3/H4 tetramer and may allow a graded transfer of such information to new tetramers containing the exchange-specific variant H3.3.

Finally, additional understanding of the mechanism by which pol II transcribes through nucleosomes is limited by the complexity of the process in vivo. Clearly elongation factors play a role in facilitating histone transfer and/or displacement from the template and it is likely that histone chaperones play a role in the elongation mechanism (Saunders et al. 2003; Levchenko et al. 2005). Moreover, transcription in vivo is likely to involve significant torsional stress within the DNA and recent experiments suggest that this stress will play a significant role in the displacement or retention of specific histones on the template. Therefore in vitro experiments will have to recapitulate the torsionally strained context in which polymerases encroach upon nucleosomes as well as include the appropriate histone chaperones and elongation factors.

Acknowledgements This work was supported by NIH grant GM52426 and NSF grant MCB-0317935. We thank Drs. Anthony Annunziato and Vasily Studitsky for a critical reading of the manuscript.

References

Adkins MW, Howar SR, Tyler JK (2004) Chromatin disassembly mediated by the histone chaperone Asf1 is essential for transcriptional activation of the yeast PHO5 and PHO8 genes. Mol Cell 14:657–666

Ahmad K, Henikoff S (2002) The histone variant H3.3 marks active chromatin by replication-independent nucleosome assembly. Mol Cell 9:1191–1200

Alilat M, Sivolob A, Revet B, Prunell A (1999) Nucleosome dynamics. Protein and DNA contributions in the chiral transition of the tetrasome, the histone (H3-H4)2 tetramer-DNA particle. J Mol Biol 291:815–841

Allfrey VG, Faulkner R, Mirsky AE (1964) Acetylation and Methylation of Histones and Their Possible Role in the Regulation of Rna Synthesis. Proc Natl Acad Sci USA 51:786–794

Annunziato AA, Frado L-LY, Seale RL, Woodcock CLF (1988) Treatment with sodium butyrate inhibits the complete condensation of interphase chromatin. Chromosoma 96:132–138

Belotserkovskaya R, Oh S, Bondarenko VA, Orphanides G, Studitsky VM, Reinberg D (2003) FACT facilitates transcription-dependent nucleosome alteration. Science 301:1090–1093

Boeger H, Griesenbeck J, Strattan JS, Kornberg RD (2003) Nucleosomes unfold completely at a transcriptionally active promoter. Mol Cell 11:1587–1598

Boyer LA, Logie C, Bonte E, Becker PB, Wade PA, Wolffe AP, Wu C, Imbalzano AN, Peterson CL (2000) Functional delineation of three groups of the ATP-dependent family of chromatin remodeling enzymes. J Biol Chem 275:18864–18870

Brownell JE, Zhou J, Ranalli T, Kobayashi R, Edmondson DG, Roth SY, Allis CD (1996) Tetrahymena histone acetyltransferase A: a homolog to yeast Gcn5p linking histone acetylation to gene activation. Cell 84:843 851

Chang CH, Luse DS (1997) The H3/H4 tetramer blocks transcript elongation by RNA polymerase II in vitro. J Biol Chem 272:23427–23434

Clark DJ, Felsenfeld G (1992) A nucleosome core is transferred out of the path of a transcribing polymerase. Cell 71:11–22

Cosma MP (2002) Ordered recruitment: gene-specific mechanism of transcription activation. Mol Cell 10:227–236 de la Cruz X, Lois S, Sanchez-Molina S, Martinez-Balbas M (2005) Do protein motifs read the histone code? Bioessays 27:164–175

Felsenfeld G, Clark D, Studitsky V (2000) Transcription through nucleosomes. Biophys Chem 86:231–237

Giaever GN, Wang JC (1988) Supercoiling of intracellular DNA can occur in eukaryotic cells. Cell 55:849–856

Grunstein M (1997) Histone acetylation in chromatin structure and transcription. Nature 389:349–352

Hake SB, Garcia BA, Duncan EM, Kauer M, Dellaire G, Shabanowitz J, Bazett-Jones DP, Allis CD, Hunt DF (2005) Expression patterns and post-translational modifications associated with mammalian histone H3 variants. J Biol Chem

Hansen JC (2002) Conformational Dynamics of the Chromatin Fiber in Solution: Determinants, Mechanisms, and Functions. Annu Rev Biophys Biomol Struct 31:361–392

Hendzel MJ, Davie JR (1990) Nucleosomal histones of transcriptionally active/competent chromatin preferentially exchange with newly synthesized histones in quiescent chicken erythrocytes. Biochem J 271:67–73

Huang RC, Bonner J (1962) Histone, a suppressor of chromosomal RNA synthesis. Proc Natl Acad Sci USA 48:1216–1222

Izban MG, Luse DS (1991) Transcription on nucleosomal templates by RNA polymerase II in vitro: inhibition of elongation with enhancement of sequence-specific pausing. Genes Dev 5:683–696

Jackson V, Chalkley R (1985) Histone synthesis and deposition in the G1 and S phases of hepatoma tissue culture cells. Biochemistry 24:6921–6930

Jackson V, Marshall S, Chalkley R (1981) The sites of deposition of newly synthesized histone. Nucleic Acids Res 9:4563–4581

Janicki SM, Tsukamoto T, Salghetti SE, Tansey WP, Sachidanandam R, Prasanth KV, Ried T, Shav-Tal Y, Bertrand E, Singer RH, Spector DL (2004) From silencing to gene expression: real-time analysis in single cells. Cell 116:683–698

Kimura H, Cook PR (2001) Kinetics of core histones in living human cells: little exchange of H3 and H4 and some rapid exchange of H2B. J Cell Biol 153:1341–1353

Kireeva ML, Hancock B, Cremona GH, Walter W, Studitsky VM, Kashlev M (2005) Nature of the nucleosomal barrier to RNA polymerase II. Mol Cell 18:97–108

Kireeva ML, Walter W, Tchernajenko V, Bondarenko V, Kashlev M, Studitsky VM (2002) Nucleosome remodeling induced by RNA polymerase II: loss of the H2A/H2B dimer during transcription. Mol Cell 9:541–552

Knezetic JA, Luse DS (1986) The Presence of Nucleosomes on a DNA Template Prevents Initiation by RNA Polymerase II In vitro. Cell 45:95–104

Lee CK, Shibata Y, Rao B, Strahl BD, Lieb JD (2004) Evidence for nucleosome depletion at active regulatory regions genome-wide. Nat Genet 36:900–905

Levchenko V, Jackson B, Jackson V (2005) Histone release during transcription: displacement of the two H2A-H2B dimers in the nucleosome is dependent on different levels of transcription-induced positive stress. Biochemistry 44:5357–5372

Louters L, Chalkley R (1985) Exchange of histones H1, H2A, and H2B in vivo. Biochemistry 24:3080–3085

McKittrick E, Gafken PR, Ahmad K, Henikoff S (2004) Histone H3.3 is enriched in covalent modifications associated with active chromatin. Proc Natl Acad Sci USA 101:1525–1530

Mito Y, Henikoff JG, Henikoff S (2005) Genome-scale profiling of histone H3.3 replacement patterns. Nat Genet 37:1090–1097

Orphanides G, LeRoy G, Chang CH, Luse DS, Reinberg D (1998) FACT, a factor that facilitates transcript elongation through nucleosomes. Cell 92:105–116

Perry CA, Dadd CA, Allis CD, Annunziato AT (1993) Analysis of nucleosome assembly and histone exchange using antibodies specific for acetylated H4. Biochemistry 32:13605–13614

Peterson CL (2000) ATP-dependent chromatin remodeling: going mobile. FEBS Lett 476:68–72

Peterson CL, Laniel MA (2004) Histones and histone modifications. Curr Biol 14:R546–R551

Reinke H, Horz W (2003) Histones are first hyperacetylated and then lose contact with the activated PHO5 promoter. Mol Cell 11:1599–1607

Sathyanarayana UG, Freeman LA, Lee MS, Garrard WT (1999) RNA polymerase-specific nucleosome disruption by transcription in vivo. J Biol Chem 274:16431–16436

Saunders A, Werner J, Andrulis ED, Nakayama T, Hirose S, Reinberg D, Lis JT (2003) Tracking FACT and the RNA polymerase II elongation complex through chromatin in vivo. Science 301:1094–1096

Schwabish MA, Struhl K (2004) Evidence for eviction and rapid deposition of histones upon transcriptional elongation by RNA polymerase II. Mol Cell Biol 24:10111–10117

Silverman B, Mirsky AE (1973) Accessibility of DNA in chromatin to DNA polymerase and RNA polymerase. Proc Natl Acad Sci USA 70:1326–1330

Strahl BD, Allis CD (2000) The language of covalent histone modifications. Nature 403:41–45

Studitsky VM, Clark DJ, Felsenfeld G (1994) A histone octamer can step around a transcribing polymerase without leaving the template. Cell 76:371–382

Studitsky VM, Clark DJ, Felsenfeld G (1995) Overcoming a nucleosomal barrier to transcription. Cell 83:19–27

Studitsky VM, Kassavetis GA, Geiduschek EP, Felsenfeld G (1997) Mechanism of transcription through the nucleosome by eukaryotic RNA polymerase. Science 278:1960–1963

Studitsky VM, Walter W, Kireeva M, Kashlev M, Felsenfeld G (2004) Chromatin remodeling by RNA polymerases. Trends Biochem Sci 29:127–135

Thatcher TH, MacGaffey J, Bowen J, Horowitz S, Shapiro DL, Gorovsky MA (1994) Independent evolutionary origin of histone H3.3-like variants of animals and Tetrahymena. Nucleic Acids Res 22:180–186

Thiriet C, Hayes JJ (2005) Replication-independent core histone dynamics at transcriptionally active loci in vivo. Genes Dev 19:677–682

Tsao YP, Wu HY, Liu LF (1989) Transcription-driven supercoiling of DNA: direct biochemical evidence from in vitro studies. Cell 56:111–118

Wirbelauer C, Bell O, Schubeler D (2005) Variant histone H3.3 is deposited at sites of nucleosomal displacement throughout transcribed genes while active histone modifications show a promoter-proximal bias. Genes Dev 19:1761–1766

Wolffe AP, Kurumizaka H (1998) The nucleosome: a powerful regulator of transcription. Prog Nucleic Acid Res Mol Biol 61:379–422

Woodcock CL, Dimitrov S (2001) Higher-order structure of chromatin and chromosomes. Curr Opin Genet Dev 11:130–135

Wu C (1980) The 5′ ends of Drosophila heat shock genes in chromatin are hypersensitive to DNase I. Nature 286:854–860

Wu C, Wong YC, Elgin SC (1979) The chromatin structure of specific genes: II. Disruption of chromatin structure during gene activity. Cell 16:807–814

Yu L, Gorovsky MA (1997) Constitutive expression, not a particular primary sequence, is the important feature of the H3 replacement variant hv2 in Tetrahymena thermophila. Mol Cell Biol 17:6303–6310

Zhang Y, Reinberg D (2001) Transcription regulation by histone methylation: interplay between different covalent modifications of the core histone tails. Genes Dev. 15:2343–2360

Zheng C, Hayes JJ (2003) Structures and interactions of the core histone tail domains. Biopolymers 68:539–546

Results Probl Cell Differ (41)
L. Brehon: Chromatin Dynamics in Cellular Function
DOI 10.1007/007/Published online: 10 February 2006
© Springer-Verlag Berlin Heidelberg 2006

The Roles of Chromatin Remodelling Factors in Replication

Ana Neves-Costa · Patrick Varga-Weisz (✉)

Babraham Institute, Cambridge CB2 4AT, UK
Patrick.varga-weisz@bbsrc.ac.uk

Abstract Dynamic changes of chromatin structure control DNA-dependent events, including DNA replication. Along with DNA, chromatin organization must be replicated to maintain genetic and epigenetic information through cell generations. Chromatin remodelling is important for several steps in replication: determination and activation of origins of replication, replication machinery progression, chromatin assembly and DNA repair. Histone chaperones such as the FACT complex assist DNA replication within chromatin, probably by facilitating both nucleosome disassembly and reassembly. ATP-dependent nucleosome remodelling enzymes of the SWI/SNF family, in particular imitation switch (ISWI)-containing complexes, have been linked to DNA and chromatin replication. They are targeted to replication sites to facilitate DNA replication and subsequent chromatin assembly.

1
Introduction

Chromatin is the in vivo form of eukaryotic genomes that allows DNA fibers to be compacted and organized in the nucleus. Chromatin proteins limit access to DNA; therefore, structural changes in chromatin control DNA-dependent events such as transcription, replication, recombination and repair.

To condense DNA fibers, chromatin structure comprises several levels of organization, the first of which is the nucleosome: 147 bp of DNA organized around a complex of eight histone proteins (Kornberg and Lorch 1999), with a central $(H3)_2/(H4)_2$ tetramer interacting with two H2A/H2B dimers. The further organization of nucleosomes into arrays leads to higher orders of compaction with fibers 30 nm in diameter stabilized by linker histones such as H1 (see: Schalch et al. 2005, reviewed in Luger and Hansen 2005). The higher-order structure of chromatin shows that nucleosomes not only constrain the DNA molecule, but also bring sequences together and thus factors that would be apart from each other on a linear stretch of DNA. The introduction of histone variants into nucleosomes may lead to fibers with structurally and functionally different properties (reviewed in Henikoff and Ahmad 2005).

Chromatin structure is remodelled through post-translational modifications of histones and by factors that modulate the interaction of chromatin proteins with DNA. The latter can be achieved in an energy-independent way as with histone chaperones or by ATP-dependent chromatin remodelling factors. Histone modifications include reversible acetylation, methylation, phosphorylation and ubiquitination of specific residues and are usually found clustered in the histone tails that protrude from the nucleosome core body (Peterson and Laniel 2004). Specific histone modifications have been shown to reflect transcriptional activity of chromatin regions. Most prominently, histone acetylation is usually associated with transcribed sites. Importantly, the patterns of histone modifications control DNA accessibility and interactions of proteins with the nucleosome and the chromatin fiber (Cosgrove and Wolberger 2005).

SWI/SNF-type proteins are the motors in chromatin remodelling complexes that use the energy gained by ATP-hydrolysis to change the structure of nucleoprotein complexes. It is thought that they achieve this by tracking along the DNA while being attached to the protein component of the nucleoprotein complex (reviewed in Becker 2005). Several SWI/SNF-family members have been shown to target the nucleosome, and may drive the movement of nucleosomes along the DNA fiber or their partial or complete disruption (Becker and Horz 2002; Lusser and Kadonaga 2003). Members of this family play roles in transcription, replication, recombination and repair. These factors share an ATPase domain of approximately 400 amino acid residues, initially identified in the *S. cerevisiae* transcriptional regulator Swi2/Snf2 (Eisen et al. 1995). Nucleosome remodelling ATPases can be grouped into the SWI2/SNF2, ISWI, INO80/SWR1 and CHD/Mi-2 subfamilies that exhibit distinct biochemical activities (Becker and Horz 2002; Eberharter and Becker 2004; Lusser and Kadonaga 2003). Members of a subfamily, e.g., ISWI, can form several functionally distinct complexes within a cell (Fig. 1, see below).

Dynamic structural changes in chromatin occur during DNA replication: Ahead of the replication fork, there is nucleosome disassembly rendering DNA accessible to a series of factors involved in the recognition of origins of replication, the unwinding and separation of the DNA strands and the synthesis of new nucleotide chains. Following the replication fork, chromatin is reassembled using a complement of "old" histones and of newly synthesized histones (reviewed in Annunziato 2005). Chromatin itself carries epigenetic information, as histones contain a multitude of modification marks. These marks, together with DNA methylation contribute to the establishment of expression patterns that define cell lineages and are potentially heritable through multiple cell generations. Epigenetic states must be maintained during or following DNA replication. Therefore the organization of chromatin fibers, including histone modification patterns, has to be copied to the daughter cells (Aligianni and Varga-Weisz 2005; Ehrenhofer-Murray 2004; McNairn and Gilbert 2003).

In the budding yeast *Saccharomyces cerevisiae* (*S. cerevisiae*) DNA replication initiates from sequence-specific origins of replication, which are bound by the origin recognition complex (ORC). ORC, together with Cdc6p and

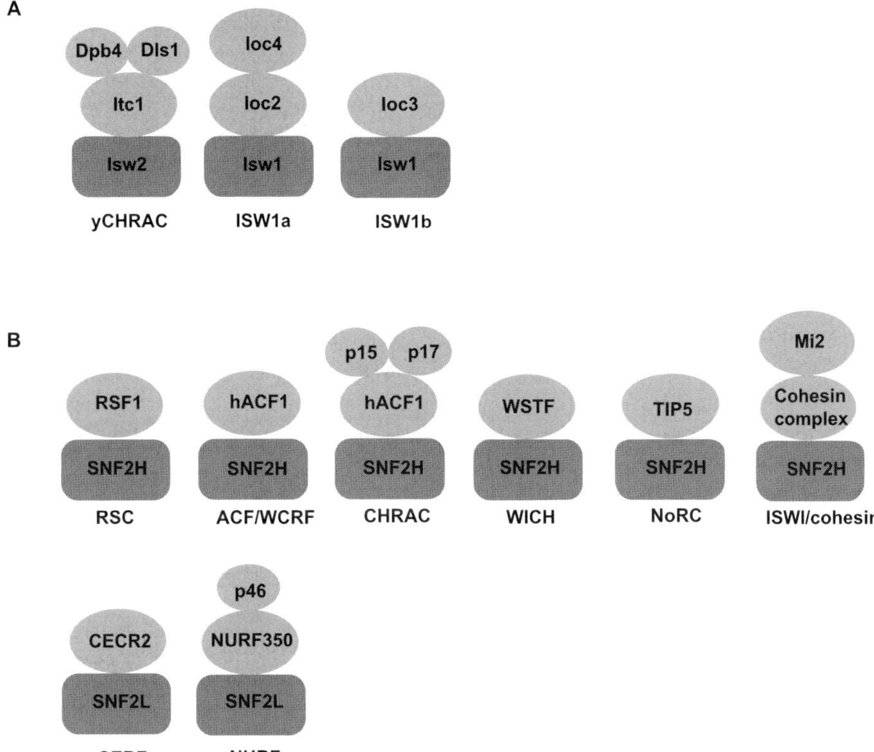

Fig. 1 *S. cerevisiae* and human ISWI complexes. **a** In *S. cerevisiae*, three ISWI complexes are described (Iida and Araki 2004; Tsukiyama et al. 1999; Vary et al. 2003). **b** In mammalian cells ISWI complexes contain ISWI-isoform SNF2H or SNF2L, (Dirscherl and Krebs 2004; Banting et al. 2005)

a hexamer of minichromosome maintenance (MCM) proteins, forms the pre-replicative complex (preRC) in the G1 phase of the cell cycle. During the S-phase, a cascade of events leads to the firing of DNA replication from the origins and processive DNA replication (for a review of DNA replication in eukaryotes: Bell and Dutta 2002). Unlike budding yeast, higher eukaryotes do not have defined sequences of origins of replication. Instead, it has been proposed that they are determined by specific contexts that may depend on particular DNA and chromatin properties (Cvetic and Walter 2005). Replication of genomes follows a defined program, whereby certain parts of the genome replicate early, and others, usually those associated with condensed chromatin structures, replicate late. Therefore, chromatin structure influences the timing of replication of specific sequences (reviewed in Donaldson 2005; McNairn and Gilbert 2003). Here we discuss the role of chromatin remodelling proteins in replication and focus particularly on energy-dependent chromatin remodelling activities.

2
Histone Modifications and DNA Replication

Histone modifications contribute to the establishment and inheritance of chromatin states. During replication, histone modifications in the nascent chromatin have to be regulated for epigenetic inheritance (Aligianni and Varga-Weisz 2005; Ehrenhofer-Murray 2004; McNairn and Gilbert 2003). Newly synthesized histones H3 and H4 are acetylated at specific lysine residues, and this form is incorporated into chromatin during replication (Annunziato 2005). However, histone underacetylation is a characteristic feature of heterochromatic regions (see Wiren et al. 2005) and to maintain heterochromatin, these modifications must subsequently be removed (Taddei et al. 1999, 2001). Histone deacetylases catalyze the removal of acetyl marks from histones and are required for heterochromatic gene silencing (see Wiren et al. 2005; Yamada et al. 2005). The importance of histone deacetylation during replication may relate to the finding that the proliferating cell nuclear antigen (PCNA), which associates with replication forks, and the clamp loader that facilitates the loading of PCNA onto DNA, interact with histone deacetylases (Anderson and Perkins 2002; Milutinovic et al. 2002).

After DNA replication, histone acetylation must be specifically reintroduced into chromatin to assure the establishment of localized transcriptional activity. Histone acetyltransferases associate with PCNA and other replication factors (Hasan et al. 2001; Meijsing and Ehrenhofer-Murray 2001; Osada et al. 2001). Histone modifications also relate to the timing of replication. Different methylation states of lysine 9 of histone H3 (mono-, di- or trimethylation) occupy distinct nuclear domains in mammalian cells. This organization correlates with the timing of origin firing during the S-phase (Wu et al. 2005). However, the relation is not fully understood, and this histone methylation per se does not seem to determine the firing of origins throughout the genome (Wu et al. 2005). Histone acetylation impacts on replication timing: Deletion of histone deacetylase Rpd3 in yeast or targeting of histone acetyltransferases to late replicating origins shifts their timing to early replication (Aparicio et al. 2004; Vogelauer et al. 2002). In summary, histone-modifying activities have important roles in the propagation of chromatin, but the mechanisms by which they regulate chromatin replication are largely unknown.

3
Histone Chaperones and DNA Replication

Histone chaperones prevent the uncontrolled association of histones with DNA and mediate their specific incorporation into or release from nucleosomes (Loyola and Almouzni 2004). Several histone chaperones are tightly connected to the replication process, delivering histones for chromatin assembly. CAF-1 (chromatin assembly factor-1) is such a histone chaperone: it binds to newly synthesized histones H3 and H4 and mediates their incorporation into DNA during replication and repair. Other histone chaperones are ASF-1 (anti-silencing factor-1) and NAP-1 (nucleosome assembly protein-1). The roles of these proteins in chromatin replication have been summarized in recent reviews (Akey and Luger 2003; Loyola and Almouzni 2004). FACT (facilitates chromatin transcription) is another factor with histone chaperone activity that has a role in DNA replication. It is a heterodimeric complex in mammalian cells and was named after its ability to promote transcription from chromatin templates (Orphanides et al. 1998). The FACT complex is highly conserved and has been purified from *S. cerevisiae*—also known as CP (Cdc68, Pob3) or SPN (Spt16/Cdc68, Pob3, Nhp6)— (Brewster et al. 1998; Formosa et al. 2001; Wittmeyer et al. 1999), *Xenopus*—also known as DUF (DNA unwinding factor, Okuhara et al. 1999) and human cells (Orphanides et al. 1999). FACT forms a stable complex with nucleosomes, binding to the histone H2A/H2B dimer, and is chromatin-associated in vivo (Formosa et al. 2001; Orphanides et al. 1999; Wittmeyer et al. 1999). Incorporation of FACT into chromatin leads to increased DNA accessibility (Rhoades et al. 2004; Seo et al. 2003). The histone chaperone function of FACT is responsible for weakening the interactions between H2A/H2B dimers and H3/H4 tetramers, destabilizing chromatin, but this function also mediates the deposition of histones onto DNA (Belotserkovskaya et al. 2003).

FACT promotes the binding of TBP-TFIIA complexes to nucleosomes, facilitating transcription initiation (Biswas et al. 2005), and counteracts the chromatin-induced block to transcription elongation (Orphanides et al. 1998). In addition to its role in transcription (Biswas et al. 2005, reviewed in Belotserkovskaya et al. 2004), FACT is implicated in DNA replication. Immunodepletion of FACT impaired the capacity of *Xenopus* egg extracts to replicate chromatin (Okuhara et al. 1999). In *S. cerevisiae*, FACT subunits directly interact with DNA polymerase α (Pol α) and this association is needed for proper S-phase progression (Wittmeyer et al. 1999; Zhou and Wang 2004). Moreover, mutations in the essential FACT subunits cause an increased sensitivity to hydroxyurea, a drug that leads to the depletion of the dNTP pool. This points to a role of FACT in antagonizing replication stress (Formosa et al. 2001; O'Donnell et al. 2004; Schlesinger and Formosa 2000). FACT may associate with catalytic activities in order to modulate chromatin structure, as suggested by the interaction with the *Xenopus* ATPase p97 (Yamada et al. 2000) and human kinase Nek9 (Tan and Lee 2004). The capacity of FACT to change DNA topology in the presence of topoisomerase I may assist the complex in its chromatin remodelling activity (Okuhara et al. 1999). In conclusion, FACT may be a key factor in facilitating DNA replication in chromatin by mediating nucleosome dynamics (Fig. 2). Its link to other chromatin remodelling activities merits further analysis.

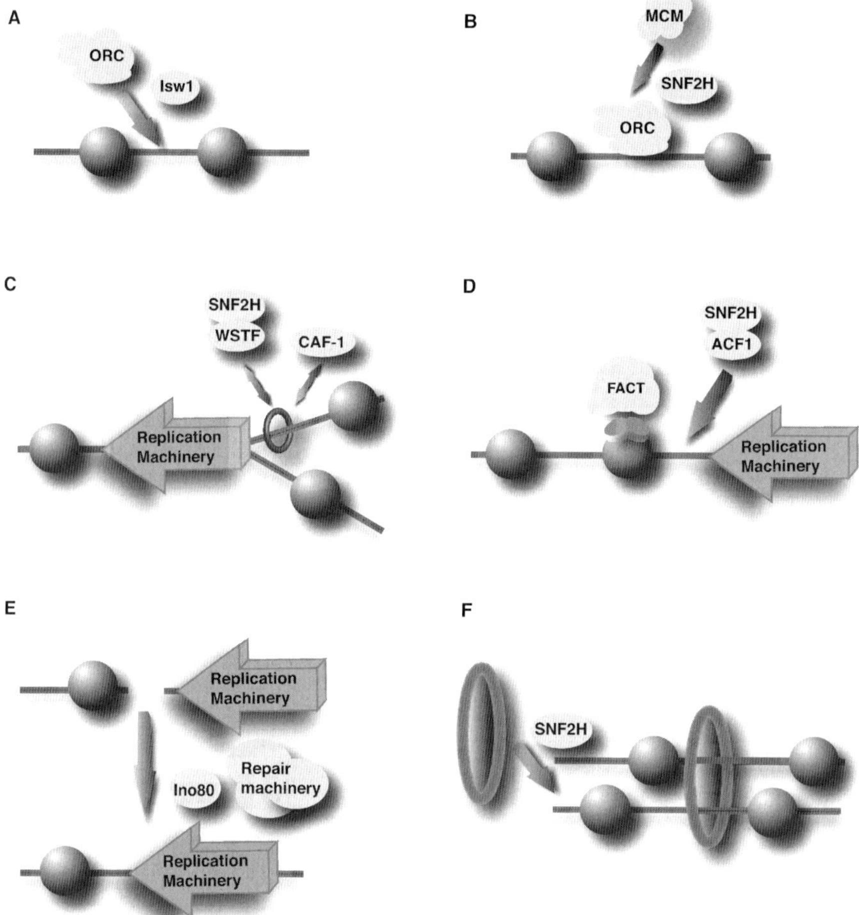

Fig. 2 Possible roles of chromatin remodelling factors in replication. **a** A genetic link between remodelling factors, such as Isw1, and origin recognition complex (ORC) components may indicate that remodellers regulate ORC function in chromatin (Suter et al. 2004). **b** Remodelling factors regulate replication firing as it has been shown that SNF2H at an origin of replication promotes chromatin binding of MCM (Zhou et al. 2005). **c** Chromatin remodelling factors such as the WSTF-SNF2H complex and histone chaperone CAF-1 are targeted by PCNA for the replication of chromatin structures (Poot et al. 2004; Akey and Luger 2003; Loyola and Almouzni 2004). **d** Histone chaperones such as FACT and ATP-dependent remodeller ACF1-SNF2H may promote DNA replication by increasing accessibility in chromatin for the replication machinery. **e** ATP-dependent chromatin remodelling factors such as Ino80 are involved in DNA repair (Morrison et al. 2004; van Attikum et al. 2004) and such activity is likely important during DNA replication, because DNA lesions have to be overcome in this process. **f** ATP-dependent chromatin remodelling factors such as SNF2H may facilitate the binding of cohesin to chromatin during replication (Hakimi et al. 2002)

4
ATP-Dependent Remodelling Factors
and Chromatin Dynamics in DNA Replication

4.1
Energy-Dependent Chromatin Remodellers Have Roles in DNA Repair

The study of the yeast ATP-dependent chromatin remodelling factors provides a framework for our understanding of the diverse functions of these proteins in chromatin dynamics. Using online databases, we find 17 SWI/SNF-type proteins in budding and 20 in the fission yeast *Schizosaccharomyces pombe* (*S. pombe*). Many SWI/SNF-type factors have been linked to DNA repair. DNA replication and DNA repair are functionally linked processes, because the cell needs nucleotide polymerization in repair and needs to overcome DNA lesions during DNA replication. Therefore, it is not unlikely that SWI/SNF-family members that have been linked to repair also have a role in DNA replication.

Rad54 is a SWI/SNF-type factor involved in homologous recombination (HR, reviewed in Krogh and Symington 2004; Tan et al. 2003). HR is a mechanism used to repair DNA double- or single-strand breaks, when a replication fork collapses or stalls, for example because it encounters a break in the DNA. This mechanism uses the homologous DNA—usually from the sister chromatid—as a template to repair lesions (Kuzminov 2001). Cells containing an ATPase-activity deficient mutant of Rad54 have increased sensitivity to DNA-damaging agents and reduced recombination (Smirnova et al. 2004). The Rad54 ATPase function mediates nucleosome remodelling in vitro: it increases nucleosomal DNA accessibility (Alexiadis et al. 2004; Jaskelioff et al. 2003) and mediates ATP-dependent nucleosome sliding and disruption (Alexeev et al. 2003, reviewed in Tan et al. 2003). These data suggest that recombination requires nucleosome remodelling to overcome the reduced accessibility to DNA due to the chromatin at the recombination sites. However, it is not clear if Rad54 specifically targets nucleosomes or has a more general capacity to remodel nucleoprotein complexes for recombination.

Other nucleosome remodelling factors have been linked to double strand DNA break repair by HR, including SWI2/SNF2 (containing the Swi2/Snf2 ATPase), RSC (containing the Sth1 ATPase) and INO80 (containing the Ino80 ATPase; Chai et al. 2005; Morrison and Shen 2005; van Attikum and Gasser 2005). In vivo, the INO80 complex is recruited to sites of double-strand breaks and facilitates DNA repair, presumably by remodelling nucleosomes to allow access of the repair machinery (Fritsch et al. 2004; Morrison et al. 2004; van Attikum et al. 2004, Fig. 2). A role in DNA repair has been suggested for another *S. cerevisiae* complex, SWR1, containing the SWI/SNF factor Swr1 (Morrison and Shen 2005). SWR1 mediates ATP-dependent deposition of the histone variant H2AZ into chromatin (Krogan et al. 2003; Mizuguchi et al. 2004; van Attikum and Gasser 2005). The mechanism could involve exchange of modified histones with unmodified ones after DNA replication, as it is known that DNA lesions induce specific histone modifications (reviewed in Moore and Krebs 2004). A similar role is known for the *Drosophila* chromatin modifying complex TIP60. In this complex the histone acetyltransferase Tip60 and the ATPase Domino interact for selective histone variant exchange at DNA lesions (Kusch et al. 2004; van Attikum and Gasser 2005).

Rad5, Rad16 and Rad26 from *S. cerevisiae* are SWI/SNF-family members linked to DNA repair, but their precise involvement in chromatin remodelling is not clear. However, the human homologue of Rad26, Cockayne syndrome B factor, has been shown to mediate ATP-dependent nucleosome remodelling in vitro (Citterio et al. 2000).

4.2
ISWI Complexes Facilitate DNA Replication in Chromatin

The ISWI (imitation switch) complexes promote sliding of nucleosomes rather than nucleosome disruption (Dirscherl and Krebs 2004; Langst and Becker 2001). This ISWI activity allows for chromatin dynamics while maintaining overall chromatin integrity. It can increase accessibility to nucleosomal DNA or promote the regular spacing of nucleosomes during chromatin assembly (Dirscherl and Krebs 2004; Langst and Becker 2001). An important characteristic of ISWI proteins is that they interact with several other proteins to form biochemically and functionally distinct nucleosome remodelling complexes that are often found in the same cell, both in yeast and mammalians (Dirscherl and Krebs 2004; Fig. 1).

In vitro studies with systems that drive replication from plasmids containing the simian virus SV40 origin of replication have proven to be useful to address chromatin replication and linked the nucleosome remodelling activity of ISWI to the initiation of DNA replication (Alexiadis et al. 1998). Recent in vivo studies using another viral origin of replication provided further insights into the mechanisms of chromatin-mediated replication control. The origin of plasmid replication, *OriP*, of the human Epstein-Barr virus was shown to be flanked by positioned nucleosomes and to associate with the mammalian ISWI isoform SNF2H in a cell cycle-dependent way (Zhou et al. 2005). The greatest enrichment of SNF2H at *OriP* was found in the G1 phase, promoting an increase in DNA accessibility and binding of the MCM complex (Zhou et al. 2005). Depletion of SNF2H from human cells causes a delay in the progression of DNA replication during S-phase (Collins et al. 2002; Zhou et al. 2005). Because this delay was no longer observed after decondensation of heterochromatin with the drug azadeoxycytidine, it was suggested that SNF2H has a role in counteracting heterochromatin (Collins et al. 2002; Fig. 2). SNF2H associates with ACF1 (ATP-utilizing chromatin assembly and remodelling factor), which is specifically recruited to replicating pericentromeric heterochromatin. ACF1 depletion also delays the progression through late S-phase and this effect can also be counteracted by azadeoxycytidine treatment (Collins et al. 2002). These observations suggest that an ACF1-SNF2H complex is involved in facilitating DNA replication in heterochromatic regions.

4.3
ISWI Complexes have Roles in the Replication of Chromatin Structures

In *S. cerevisiae*, the ISWI complex yCHRAC that is related to the *Drosophila* and mammalian ACF1-containing CHRAC (Chromatin Accessibility Complex), inhibits heterochromatin formation at telomeres: When integrated in the vicinity of telomeres, reporter genes usually become silent due to the heterochromatic arrangement close to telomeres. However, yCHRAC counteracts the association of the heterochromatin-promoting Sir3 protein with the DNA at the telomeres (Iida and Araki 2004). In contrast to yCHRAC, the catalytic subunit of Pol ε promotes transcriptional silencing and heterochromatin maintenance at this locus (Iida and Araki 2004). Similar effects of yCHRAC on chromatin structure were observed close to the silent mating type loci (Tackett et al. 2005). CHRAC and Pol ε share a subunit, Dpb4, which suggests a functional competition between the complexes (Iida and Araki 2004). Whether yCHRAC participates in the maintenance of chromatin organization and of epigenetic patterns during replication has yet to be examined.

Another way by which a chromatin remodelling factor may affect chromatin replication is by involvement in sister chromatid cohesion. Cohesin is a protein complex that holds sister chromatids together for correct chromosome segregation in mitosis, and this cohesion is established during DNA replication. Insights into a possible function for SNF2H in sister chromatid cohesion came from the purification of a remodelling complex from human cells that contains SNF2H and four subunits of the core-cohesin complex (Hakimi et al. 2002). Binding sites of SNF2H and the cohesin subunit hRad21 were determined by chromatin immunoprecipitation and found to overlap in several chromosomes. The association of the SNF2H/cohesin complex with chromatin appears be of functional relevance, as it is dependent on the ATPase activity of SNF2H (Hakimi et al. 2002). The SNF2H ATP-dependent remodelling activity may be required for the maintenance of a chromatin state that favors sister chromatid cohesion (Fig. 2). Budding yeast remodelling factor RSC has also been implicated in facilitating cohesin binding (Huang et al. 2004; Baetz et al. 2004; reviewed in Riedel et al. 2004).

4.4
ISWI Complexes Target Replication Sites

The involvement of ISWI complexes in replication is underlined by the observation that these factors target replication foci (Bozhenok et al. 2002; Collins et al. 2002; Poot et al. 2004). WICH, a complex between ISWI and the Williams syndrome transcription factor, WSTF, appears to target replication sites by interacting with PCNA (Poot et al. 2004; Fig. 2). PCNA encircles the DNA molecule and is a binding platform for DNA polymerases and other factors involved in DNA and chromatin replication (Maga and Hubscher 2003). WSTF depletion causes condensation of nascent chromatin, as well as a cellular increase in heterochromatin markers heterochromatin protein 1 α and β (HP1α and HP1β, Poot et al. 2004, 2005). WICH may be involved in preventing aberrant formation of heterochromatin by maintaining an open chromatin structure after DNA replication. This may facilitate rebinding to the newly replicated chromatin of factors that have been evicted from the parental chromatin during the replication process (Poot et al. 2005). An important role of SNF2H in cell proliferation is underscored by the fact that this protein is essential in proliferating mammalian cells and is upregulated in rapidly proliferating cells (Lazzaro and Picketts 2001; Stopka and Skoultchi 2003).

A potential link between ATP-dependent nucleosome remodellers and DNA replication was uncovered in a genetic screen in *S. cerevisiae* that identified two SWI/SNF factors, Isw1 (a yeast ISWI-isoform) and Fun30, as genetic partners of the ORC (Suter et al. 2004). The combination of viable mutations in the Orc2 and Orc5 subunits of ORC, with the viable deletion of either Fun30 or Isw1, caused defective growth or lethality. These genetic interactions may reflect functional interactions, indicating that these chromatin remodelling factors might have a role in facilitating origin binding by ORC (Fig. 2). In line with the idea that remodelling complexes facilitate ORC function, an analysis in budding yeast indicated a role for the SWI/SNF complex in the function of some origins of replication (Flanagan and Peterson 1999).

ISWI complex ACF has been shown to be involved in chromatin assembly in the fly consistent with their demonstrated role in the assembly of regular nucleosome arrays in vitro (Fyodorov et al. 2004). In mammalian cells, an ACF1-SNF2H and the SNF2H-containing NoRC complex have been linked to gene silencing (Santoro and Grummt 2005; Yasui et al. 2002). The formation of repressive chromatin linked to silencing of specific genes is also performed by yCHRAC (Goldmark et al. 2000; McConnell et al. 2004). We may conclude that the major activity of ISWI complexes is to facilitate dynamic transitions in chromatin, and these may lead either to gene activation or gene silencing, depending on the chromatin context and the presence of other chromatin determinants.

Despite the fact that ISWI is conserved between *S. cerevisiae* and human, fission yeast (*S. pombe*) does not have an obvious ISWI homologue. Other related proteins may function as substitutes, and Hrp1 is a possible candidate. Recent investigations on Hrp1 provided evidence that it has a role in promoting chromatin assembly during DNA replication: Hrp1 appears to maintain silencing at the centromeres by promoting the incorporation of the histone H3 variant Cnp1 into nucleosomes during replication-coupled chromatin assembly (Walfridsson et al. 2005).

5
Conclusion

The tight regulation of replication events is used by cells to maintain their genomic and epigenomic content over cell divisions. Chromatin-remodelling complexes are players in this regulation and seem to participate in all stages of the replication process (Fig. 2). In particular, ISWI complexes have diverse roles at replication sites. Yet this field is at its beginning and much work is needed to shed light on the mechanisms of chromatin replication. Future studies concerning chromatin dynamics at replication sites will have to address: (1) which factors and complexes are required for the different levels of regulation of replication; (2) how the dynamic recruitment of factors to replication sites is achieved; (3) how chromatin structure and domains are inherited through replication; (4) how higher-order structures of chromatin are affected during replication and re-established after it.

S. pombe and *S. cerevisiae* are prime model organisms for further functional studies concerning these questions. These organisms share various important chromatin features with higher eukaryotes in a complementary way: Whereas *S. pombe* has heterochromatin complexes closely related to their counterparts in higher eukaryotes, but not found in *S. cerevisiae* (Huang 2002), *S. cerevisiae* shares remodelling activities with higher eukaryotes, which are absent in *S. pombe*. Even these simple organisms contain a plethora of SWI/SNF-type ATPases, several of which are only poorly characterized. Novel perspectives on DNA and chromatin replication may arise from studies of some of these less characterized factors.

Acknowledgements We thank Tom Sexton for constructive criticism of this manuscript. ANC is funded by Fundação para a Ciência e Tecnologia. Work in the Varga-Weisz laboratory is funded by the Biotechnology and Biosciences Research Council, the Association for International Cancer Research, St. Andrews (AICR) and the European Network of Excellence (NoE).

References

Akey CW, Luger K (2003) Histone chaperones and nucleosome assembly. Curr Opin Struct Biol 13:6–14

Alexeev A, Mazin A, Kowalczykowski SC (2003) Rad54 protein possesses chromatin-remodeling activity stimulated by the Rad51-ssDNA nucleoprotein filament. Nat Struct Biol 10:182–186

Alexiadis V, Lusser A, Kadonaga JT (2004) A conserved N-terminal motif in Rad54 is important for chromatin remodeling and homologous strand pairing. J Biol Chem 279:27824–27829

Alexiadis V, Varga-Weisz PD, Bonte E, Becker PB, Gruss C (1998) In vitro chromatin remodelling by chromatin accessibility complex (CHRAC) at the SV40 origin of DNA replication. Embo J 17:3428–3438

Aligianni S, Varga-Weisz P (2005) Chromatin-remodelling factors and the maintenance of transcriptional states through DNA replication. Biochem Soc Symp 73:97–108

Anderson LA, Perkins ND (2002) The large subunit of replication factor C interacts with the histone deacetylase, HDAC1. J Biol Chem 277:29550–29554

Annunziato AT (2005) Split decision: what happens to nucleosomes during DNA replication? J Biol Chem 280:12065–12068

Aparicio JG, Viggiani CJ, Gibson DG, Aparicio OM (2004) The Rpd3-Sin3 histone deacetylase regulates replication timing and enables intra-S origin control in *Saccharomyces cerevisiae*. Mol Cell Biol 24:4769–4780

Baetz KK, Krogan NJ, Emili A, Greenblatt J, Hieter P (2004) The ctf13–30/CTF13 genomic haploinsufficiency modifier screen identifies the yeast chromatin remodeling complex RSC, which is required for the establishment of sister chromatid cohesion. Mol Cell Biol 24:1232–1244

Banting GS, Barak O, Ames TM, Burnham AC, Kardel MD, Cooch NS, Davidson CE, Godbout R, McDermid HE, Shiekhattar R (2005) CECR2, a protein involved in neurulation, forms a novel chromatin remodeling complex with SNF2L. Hum Mol Genet 14:513–524

Barak O, Lazzaro MA, Lane WS, Speicher DW, Picketts DJ, Shiekhattar R (2003) Isolation of human NURF: a regulator of Engrailed gene expression. Embo J 22:6089–6100

Becker PB (2005) Nucleosome remodelers on track. Nat Struct Mol Biol 12:732–733

Becker PB, Horz W (2002) ATP-dependent nucleosome remodeling. Annu Rev Biochem 71:247–273

Bell SP, Dutta A (2002) DNA replication in eukaryotic cells. Annu Rev Biochem 71:333–374

Belotserkovskaya R, Oh S, Bondarenko VA, Orphanides G, Studitsky VM, Reinberg D (2003) FACT facilitates transcription-dependent nucleosome alteration. Science 301:1090–1093

Belotserkovskaya R, Saunders A, Lis JT, Reinberg D (2004) Transcription through chromatin: understanding a complex FACT. Biochim Biophys Acta 1677:87–99

Biswas D, Yu Y, Prall M, Formosa T, Stillman DJ (2005) The yeast FACT complex has a role in transcriptional initiation. Mol Cell Biol 25:5812–5822

Bochar DA, Savard J, Wang W, Lafleur DW, Moore P, Cote J, Shiekhattar R (2000) A family of chromatin remodeling factors related to Williams syndrome transcription factor. Proc Natl Acad Sci USA 97:1038–1043

Bozhenok L, Wade PA, Varga-Weisz P (2002) WSTF-ISWI chromatin remodeling complex targets heterochromatic replication foci. Embo J 21:2231–2241

Brewster NK, Johnston GC, Singer RA (1998) Characterization of the CP complex, an abundant dimer of Cdc68 and Pob3 proteins that regulates yeast transcriptional activation and chromatin repression. J Biol Chem 273:21972–21979

Chai B, Huang J, Cairns BR, Laurent BC (2005) Distinct roles for the RSC and Swi/Snf ATP-dependent chromatin remodelers in DNA double-strand break repair. Genes Dev 19:1656–1661

Citterio E, Van Den Boom V, Schnitzler G, Kanaar R, Bonte E, Kingston RE, Hoeijmakers JH, Vermeulen W (2000) ATP-dependent chromatin remodeling by the Cockayne syndrome B DNA repair-transcription-coupling factor. Mol Cell Biol 20:7643–7653

Collins N, Poot RA, Kukimoto I, Garcia-Jimenez C, Dellaire G, Varga-Weisz PD (2002) An ACF1-ISWI chromatin-remodeling complex is required for DNA replication through heterochromatin. Nat Genet 32:627–632

Cosgrove MS, Wolberger C (2005) How does the histone code work? Biochem Cell Biol 83:468–476

Cvetic C, Walter JC (2005) Eukaryotic origins of DNA replication: could you please be more specific? Semin Cell Dev Biol 16:343–353

Dirscherl SS, Krebs JE (2004) Functional diversity of ISWI complexes. Biochem Cell Biol 82:482–489

Donaldson AD (2005) Shaping time: chromatin structure and the DNA replication programme. Trends Genet 21:444–449

Eberharter A, Becker PB (2004) ATP-dependent nucleosome remodelling: factors and functions. J Cell Sci 117:3707–3711

Ehrenhofer-Murray AE (2004) Chromatin dynamics at DNA replication, transcription and repair. Eur J Biochem 271:2335–2349

Eisen JA, Sweder KS, Hanawalt PC (1995) Evolution of the SNF2 family of proteins: subfamilies with distinct sequences and functions. Nucleic Acids Res 23:2715–2723

Flanagan JF, Peterson CL (1999) A role for the yeast SWI/SNF complex in DNA replication. Nucleic Acids Res 27:2022–2028

Formosa T, Eriksson P, Wittmeyer J, Ginn J, Yu Y, Stillman DJ (2001) Spt16-Pob3 and the HMG protein Nhp6 combine to form the nucleosome-binding factor SPN. Embo J 20:3506–3517

Fritsch O, Benvenuto G, Bowler C, Molinier J, Hohn B (2004) The INO80 protein controls homologous recombination in *Arabidopsis thaliana*. Mol Cell 16:479–485

Fyodorov DV, Blower MD, Karpen GH, Kadonaga JT (2004) Acf1 confers unique activities to ACF/CHRAC and promotes the formation rather than disruption of chromatin in vivo. Genes Dev 18:170–183

Goldmark JP, Fazzio TG, Estep PW, Church GM, Tsukiyama T (2000) The Isw2 chromatin remodeling complex represses early meiotic genes upon recruitment by Ume6p. Cell 103:423–433

Guschin D, Geiman TM, Kikyo N, Tremethick DJ, Wolffe AP, Wade PA (2000) Multiple ISWI ATPase complexes from *Xenopus laevis*. Functional conservation of an ACF/CHRAC homolog. J Biol Chem 275:35248–35255

Hakimi MA, Bochar DA, Schmiesing JA, Dong Y, Barak OG, Speicher DW, Yokomori K, Shiekhattar R (2002) A chromatin remodelling complex that loads cohesin onto human chromosomes. Nature 418:994–998

Hasan S, Stucki M, Hassa PO, Imhof R, Gehrig P, Hunziker P, Hubscher U, Hottiger MO (2001) Regulation of human flap endonuclease-1 activity by acetylation through the transcriptional coactivator p300. Mol Cell 7:1221–1231

Henikoff S, Ahmad K (2005) Assembly of Variant Histones into Chromatin. Annu Rev Cell Dev Biol 21:133–153

Huang J, Hsu JM, Laurent BC (2004) The RSC nucleosome-remodeling complex is required for Cohesin's association with chromosome arms. Mol Cell 13:739–750

Huang Y (2002) Transcriptional silencing in *Saccharomyces cerevisiae* and *Schizosaccharomyces pombe*. Nucleic Acids Res 30:1465–1482

Iida T, Araki H (2004) Noncompetitive counteractions of DNA polymerase epsilon and ISW2/yCHRAC for epigenetic inheritance of telomere position effect in *Saccharomyces cerevisiae*. Mol Cell Biol 24:217–227

Ito T, Bulger M, Pazin MJ, Kobayashi R, Kadonaga JT (1997) ACF, an ISWI-containing and ATP-utilizing chromatin assembly and remodeling factor. Cell 90:145–155

Jaskelioff M, Van Komen S, Krebs JE, Sung P, Peterson CL (2003) Rad54p is a chromatin remodeling enzyme required for heteroduplex DNA joint formation with chromatin. J Biol Chem 278:9212–9218

Kornberg RD, Lorch Y (1999) Twenty-five years of the nucleosome, fundamental particle of the eukaryote chromosome. Cell 98:285–294

Krogan NJ, Keogh MC, Datta N, Sawa C, Ryan OW, Ding H, Haw RA, Pootoolal J, Tong A, Canadien V, Richards DP, Wu X, Emili A, Hughes TR, Buratowski S, Greenblatt JF (2003) A Snf2 family ATPase complex required for recruitment of the histone H2A variant Htz1. Mol Cell 12:1565–1576

Krogh BO, Symington LS (2004) Recombination proteins in yeast. Annu Rev Genet 38:233–271

Kusch T, Florens L, Macdonald WH, Swanson SK, Glaser RL, Yates JR III, Abmayr SM, Washburn MP, Workman JL (2004) Acetylation by Tip60 is required for selective histone variant exchange at DNA lesions. Science 306:2084–2087

Kuzminov A (2001) DNA replication meets genetic exchange: chromosomal damage and its repair by homologous recombination. Proc Natl Acad Sci USA 98:8461–8468

Langst G, Becker PB (2001) Nucleosome mobilization and positioning by ISWI-containing chromatin-remodeling factors. J Cell Sci 114:2561–2568

Lazzaro MA, Picketts DJ (2001) Cloning and characterization of the murine Imitation Switch (ISWI) genes: differential expression patterns suggest distinct developmental roles for Snf2h and Snf2l. J Neurochem 77:1145–1156

LeRoy G, Loyola A, Lane WS, Reinberg D (2000) Purification and characterization of a human factor that assembles and remodels chromatin. J Biol Chem 275:14787–14790

LeRoy G, Orphanides G, Lane WS, Reinberg D (1998) Requirement of RSF and FACT for transcription of chromatin templates in vitro. Science 282:1900–1904

Loyola A, Almouzni G (2004) Histone chaperones, a supporting role in the limelight. Biochim Biophys Acta 1677:3–11

Luger K, Hansen JC (2005) Nucleosome and chromatin fiber dynamics. Curr Opin Struct Biol 15:188–196

Lusser A, Kadonaga JT (2003) Chromatin remodeling by ATP-dependent molecular machines. Bioessays 25:1192–1200

MacCallum DE, Losada A, Kobayashi R, Hirano T (2002) ISWI remodeling complexes in *Xenopus* egg extracts: identification as major chromosomal components that are regulated by INCENP-aurora B. Mol Biol Cell 13:25–39

Maga G, Hubscher U (2003) Proliferating cell nuclear antigen (PCNA): a dancer with many partners. J Cell Sci 116:3051–3060

McConnell AD, Gelbart ME, Tsukiyama T (2004) Histone fold protein Dls1p is required for Isw2-dependent chromatin remodeling in vivo. Mol Cell Biol 24:2605–2613

McNairn AJ, Gilbert DM (2003) Epigenomic replication: linking epigenetics to DNA replication. Bioessays 25:647–656

Meijsing SH, Ehrenhofer-Murray AE (2001) The silencing complex SAS-I links histone acetylation to the assembly of repressed chromatin by CAF-I and Asf1 in *Saccharomyces cerevisiae*. Genes Dev 15:3169–3182

Milutinovic S, Zhuang Q, Szyf M (2002) Proliferating cell nuclear antigen associates with histone deacetylase activity, integrating DNA replication and chromatin modification. J Biol Chem 277:20974–20978

Mizuguchi G, Shen X, Landry J, Wu WH, Sen S, Wu C (2004) ATP-driven exchange of histone H2AZ variant catalyzed by SWR1 chromatin remodeling complex. Science 303:343–348

Moore JD, Krebs JE (2004) Histone modifications and DNA double-strand break repair. Biochem Cell Biol 82:446–452

Morrison AJ, Highland J, Krogan NJ, Arbel-Eden A, Greenblatt JF, Haber JE, Shen X (2004) INO80 and gamma-H2AX interaction links ATP-dependent chromatin remodeling to DNA damage repair. Cell 119:767–775

Morrison AJ, Shen X (2005) DNA repair in the context of chromatin. Cell Cycle 4:568–571

O'Donnell AF, Brewster NK, Kurniawan J, Minard LV, Johnston GC, Singer RA (2004) Domain organization of the yeast histone chaperone FACT: the conserved N-terminal domain of FACT subunit Spt16 mediates recovery from replication stress. Nucleic Acids Res 32:5894–5906

Okuhara K, Ohta K, Seo H, Shioda M, Yamada T, Tanaka Y, Dohmae N, Seyama Y, Shibata T, Murofushi H (1999) A DNA unwinding factor involved in DNA replication in cell-free extracts of Xenopus eggs. Curr Biol 9:341–350

Orphanides G, LeRoy G, Chang CH, Luse DS, Reinberg D (1998) FACT, a factor that facilitates transcript elongation through nucleosomes. Cell 92:105–116

Orphanides G, Wu WH, Lane WS, Hampsey M, Reinberg D (1999) The chromatin-specific transcription elongation factor FACT comprises human SPT16 and SSRP1 proteins. Nature 400:284–288

Osada S, Sutton A, Muster N, Brown CE, Yates JR 3rd, Sternglanz R, Workman JL (2001) The yeast SAS (something about silencing) protein complex contains a MYST-type putative acetyltransferase and functions with chromatin assembly factor ASF1. Genes Dev 15:3155–3168

Peterson CL, Laniel MA (2004) Histones and histone modifications. Curr Biol 14:R546–R551

Poot RA, Bozhenok L, van den Berg DL, Hawkes N, Varga-Weisz PD (2005) Chromatin remodeling by WSTF-ISWI at the replication site: opening a window of opportunity for epigenetic inheritance? Cell Cycle 4:543–546

Poot RA, Bozhenok L, van den Berg DL, Steffensen S, Ferreira F, Grimaldi M, Gilbert N, Ferreira J, Varga-Weisz PD (2004) The Williams syndrome transcription factor interacts with PCNA to target chromatin remodelling by ISWI to replication foci. Nat Cell Biol 6:1236–1244

Poot RA, Dellaire G, Hulsmann BB, Grimaldi MA, Corona DF, Becker PB, Bickmore WA, Varga-Weisz PD (2000) HuCHRAC, a human ISWI chromatin remodelling complex contains hACF1 and two novel histone-fold proteins. Embo J 19:3377–3387

Rhoades AR, Ruone S, Formosa T (2004) Structural features of nucleosomes reorganized by yeast FACT and its HMG box component, Nhp6. Mol Cell Biol 24:3907–3917

Riedel CG, Gregan J, Gruber S, Nasmyth K (2004) Is chromatin remodeling required to build sister-chromatid cohesion? Trends Biochem Sci 29:389–392

Santoro R, Grummt I (2005) Epigenetic mechanism of rRNA gene silencing: temporal order of NoRC-mediated histone modification, chromatin remodeling, and DNA methylation. Mol Cell Biol 25:2539–2546

Schalch T, Duda S, Sargent DF, Richmond TJ (2005) X-ray structure of a tetranucleosome and its implications for the chromatin fibre. Nature 436:138–141

Schlesinger MB, Formosa T (2000) POB3 is required for both transcription and replication in the yeast *Saccharomyces cerevisiae*. Genetics 155:1593–1606

Seo H, Okuhara K, Kurumizaka H, Yamada T, Shibata T, Ohta K, Akiyama T, Murofushi H (2003) Incorporation of DUF/FACT into chromatin enhances the accessibility of nucleosomal DNA. Biochem Biophys Res Commun 303:8–13

Smirnova M, Van Komen S, Sung P, Klein HL (2004) Effects of tumor-associated mutations on Rad54 functions. J Biol Chem 279:24081–24088

Stopka T, Skoultchi AI (2003) The ISWI ATPase Snf2h is required for early mouse development. Proc Natl Acad Sci USA 100:14097–14102

Strohner R, Nemeth A, Jansa P, Hofmann-Rohrer U, Santoro R, Langst G, Grummt I (2001) NoRC— a novel member of mammalian ISWI-containing chromatin remodeling machines. Embo J 20:4892–4900

Suter B, Tong A, Chang M, Yu L, Brown GW, Boone C, Rine J (2004) The origin recognition complex links replication, sister chromatid cohesion and transcriptional silencing in *Saccharomyces cerevisiae*. Genetics 167:579–591

Tackett AJ, Dilworth DJ, Davey MJ, O'Donnell M, Aitchison JD, Rout MP, Chait BT (2005) Proteomic and genomic characterization of chromatin complexes at a boundary. J Cell Biol 169:35–47

Taddei A, Maison C, Roche D, Almouzni G (2001) Reversible disruption of pericentric heterochromatin and centromere function by inhibiting deacetylases. Nat Cell Biol 3:114–120

Taddei A, Roche D, Sibarita JB, Turner BM, Almouzni G (1999) Duplication and maintenance of heterochromatin domains. J Cell Biol 147:1153–1166

Tan BC, Lee SC (2004) Nek9, a novel FACT-associated protein, modulates interphase progression. J Biol Chem 279:9321–9330

Tan TL, Kanaar R, Wyman C (2003) Rad54, a Jack of all trades in homologous recombination. DNA Repair (Amst) 2:787–794

Tsukiyama T, Palmer J, Landel CC, Shiloach J, Wu C (1999) Characterization of the imitation switch subfamily of ATP-dependent chromatin-remodeling factors in Saccharomyces cerevisiae. Genes Dev 13:686–697

Tsukiyama T, Wu C (1995) Purification and properties of an ATP-dependent nucleosome remodeling factor. Cell 83:1011–1020

van Attikum H, Fritsch O, Hohn B, Gasser SM (2004) Recruitment of the INO80 complex by H2A phosphorylation links ATP-dependent chromatin remodeling with DNA double-strand break repair. Cell 119:777–788

van Attikum H, Gasser SM (2005) ATP-dependent chromatin remodeling and dna double-strand break repair. Cell Cycle 4:1011–1014

Varga-Weisz PD, Wilm M, Bonte E, Dumas K, Mann M, Becker PB (1997) Chromatin-remodelling factor CHRAC contains the ATPases ISWI and topoisomerase II. Nature 388:598–602

Vary JC Jr, Gangaraju VK, Qin J, Landel CC, Kooperberg C, Bartholomew B, Tsukiyama T (2003) Yeast Isw1p forms two separable complexes in vivo. Mol Cell Biol 23:80–91

Vogelauer M, Rubbi L, Lucas I, Brewer BJ, Grunstein M (2002) Histone acetylation regulates the time of replication origin firing. Mol Cell 10:1223–1233

Walfridsson J, Bjerling P, Thalen M, Yoo EJ, Park SD, Ekwall K (2005) The CHD remodeling factor Hrp1 stimulates CENP-A loading to centromeres. Nucleic Acids Res 33:2868–2879

Wiren M, Silverstein RA, Sinha I, Walfridsson J, Lee HM, Laurenson P, Pillus L, Robyr D, Grunstein M, Ekwall K (2005) Genomewide analysis of nucleosome density histone acetylation and HDAC function in fission yeast. Embo J 24:2906–2918

Wittmeyer J, Joss L, Formosa T (1999) Spt16 and Pob3 of Saccharomyces cerevisiae form an essential, abundant heterodimer that is nuclear, chromatin-associated, and copurifies with DNA polymerase alpha. Biochemistry 38:8961–8971

Wu R, Terry AV, Singh PB, Gilbert DM (2005) Differential subnuclear localization and replication timing of histone H3 lysine 9 methylation states. Mol Biol Cell 16:2872–2881

Yamada T, Fischle W, Sugiyama T, Allis CD, Grewal SI (2005) The nucleation and maintenance of heterochromatin by a histone deacetylase in fission yeast. Mol Cell 20:173–185

Yamada T, Okuhara K, Iwamatsu A, Seo H, Ohta K, Shibata T, Murofushi H (2000) p97 ATPase, an ATPase involved in membrane fusion, interacts with DNA unwinding factor (DUF) that functions in DNA replication. FEBS Lett 466:287–291

Yasui D, Miyano M, Cai S, Varga-Weisz P, Kohwi-Shigematsu T (2002) SATB1 targets chromatin remodelling to regulate genes over long distances. Nature 419:641–645

Zhou J, Chau CM, Deng Z, Shiekhattar R, Spindler MP, Schepers A, Lieberman PM (2005) Cell cycle regulation of chromatin at an origin of DNA replication. Embo J 24:1406–1417

Zhou Y, Wang TS (2004) A coordinated temporal interplay of nucleosome reorganization factor, sister chromatin cohesion factor, and DNA polymerase alpha facilitates DNA replication. Mol Cell Biol 24:9568–9579

Results Probl Cell Differ (41)
L. Brehon: Chromatin Dynamics in Cellular Function
DOI 10.1007/008/Published online: 24 February 2006
© Springer-Verlag Berlin Heidelberg 2006

Chromatin Modifications in DNA Repair

Ashby J. Morrison · Xuetong Shen (✉)

University of Texas, M.D. Anderson Cancer Center, Department of Carcinogenesis,
Science Park Research Division, Smithville, Texas 78957, USA
snowshen@mac.com

Abstract A requirement of nuclear processes that use DNA as a substrate is the manipulation of chromatin in which the DNA is packaged. Chromatin modifications cause alterations of histones and DNA, and result in a permissive chromatin environment for these nuclear processes. Recent advances in the fields of DNA repair and chromatin reveal that both histone modifications and chromatin-remodeling complexes are essential for the repair of DNA lesions, such as DNA double strand breaks (DSBs). In particular, chromatin-modifying complexes, such as the INO80, SWR1, RSC, and SWI/SNF ATP-dependent chromatin-remodeling complexes and the NuA4 and Tip60 histone acetyltransferase complexes are implicated in DNA repair. The activity of these chromatin-modifying complexes influences the efficiency of the DNA repair process, which ultimately affects genome integrity and carcinogenesis. Thus, the process of DNA repair requires the cooperative activities of evolutionarily conserved chromatin-modifying complexes that facilitate the dynamic chromatin alterations needed during repair of DNA damage.

1
Overview of Chromatin Modifications

1.1
Introduction

In eukaryotic cells the genome is packaged into chromatin, which consists of DNA and histones. DNA transactions (such as transcription, replication, and repair) occur in the context of chromatin and require dynamic changes of chromatin structure (Fyodorov and Kadonaga 2001). However, the packaging of the eukaryotic genome in chromatin presents barriers that restrict the access of DNA to processing enzymes (Kornberg and Lorch 1999; Luger and Richmond 1998). To counteract these constraints, the eukaryotic cell uses two major strategies to modify chromatin: ATP-dependent perturbations of histone–DNA interactions catalyzed by the SWI/SNF family of ATP-dependent chromatin-remodeling complexes, and covalent modification of histones catalyzed by histone-modifying enzyme complexes, such as histone acetyltransferases (HATs).

Studies from the past decade indicate that both ATP-dependent chromatin-remodeling and histone post-translational modifications are critical for many nuclear functions, such as transcription, DNA replication, recombination and repair (Becker and Horz 2002; Kornberg and Lorch 1999; Neely and Workman 2002; Roth et al. 2001). Of all the nuclear activities, chromatin modifications have been most extensively studied in transcription (Armstrong and Emerson 1998; Kadonaga 1998; Workman and Kingston 1998). However, little is known about the link between chromatin modifications and other nuclear events, such as DNA repair. The link between chromatin and DNA repair is of particular interest since the integrity of the genome depends on the ability of cells to repair DNA damage within the context of chromatin. Failure to repair DNA can result in genome instability and contribute to carcinogenesis. However, in the past, the field of DNA repair research has mainly focused on repair processes using DNA alone as template, thus disregarding the role of chromatin in DNA repair. Notably, many recent studies on chromatin have shown that chromatin modifications play important roles in DNA repair. These studies link chromatin to DNA repair and highlight the importance of research to investigate DNA repair in the native chromatin environment.

1.2
Chromatin Modifications

Chromatin can be modified by post-translational modifications of the histone tails through acetylation, methylation, phosphorylation, and other modifications. ATP-dependent chromatin-remodeling factors also modify chromatin by altering histone–DNA interactions, such that nucleosomal DNA becomes more accessible to interacting proteins. These two major forms of chromatin modifications enable a fluid state of the chromatin in which diverse nuclear processes can efficiently occur.

ATP-dependent chromatin-remodeling complexes have been discovered in the past decade. Through in vivo and in vitro studies, the link between chromatin remodeling and transcriptional activation has become quite strong. For example, mutations in the founding member of the SWI/SNF family of genes, the *SNF2* gene, result in transcriptional and chromatin-remodeling defects of specific genes (Hirschhorn et al. 1992; Sudarsanam et al. 2000). Biochemically, the ISWI family of remodeling complexes, such as NURF, was purified based on its activity to promote transcription from a chromatin template (Tsukiyama and Wu 1995). Many ATP-dependent chromatin-remodeling complexes have since been characterized and found to assist transcription in various systems. One common feature of these complexes is the presence of a SWI2/SNF2 family core ATPase. The four main classes of remodeling complexes (SWI/SNF, ISWI, Mi2, and INO80) are classified into different subfamilies based on their subunit composition and activities (Kingston and Narlikar 1999; Shen et al. 2000). All four classes have been found to be involved in the regulation of specific genes. Current understanding of the mechanism linking chromatin remodeling and transcription is based on evidence showing that ATP-dependent chromatin-remodeling complexes are recruited to promoter regions of specific genes by gene-specific transcriptional activators (Burns and Peterson 1997). Gene activation is achieved by local disruption of chromatin structure, thereby facilitating the access of the transcription machinery. In vitro, nearly all of these complexes have been shown to remodel chromatin and assist transcription from chromatin templates. Although the role of ATP-dependent chromatin remodeling in DNA repair is much less understood, emerging studies on this topic will be discussed later.

Mechanistically, ATP-dependent chromatin-remodeling complexes utilize the energy supplied by ATP hydrolysis to reconfigure nucleosomal organization by either "sliding" the nucleosomes along DNA or by displacing or replacing histones within nucleosomes (Mizuguchi et al. 2004). Therefore, ATP-dependent chromatin remodeling is an active and direct way to modify various chromatin structures. Similar to ATP-dependent chromatin remodeling, histone modifications are carried out by various evolutionarily conserved protein complexes, most of which have been clearly implicated in transcription. Histone modifications, such as phosphorylation, acetylation, and methylation, can alter higher order chromatin structure (Tse et al. 1998a, 1998b). Histone modifications can also serve as "histone codes", which define specific associations between chromatin and its interacting partners (Strahl and Allis 2000). As discussed later in the chapter, histone modifications are also important for DNA repair, thus a DNA repair-specific "histone code" may exist. In addition, there is also growing evidence that both chromatin-remodeling and histone-modifying activities are interconnected in the process of DNA repair.

2
Histone Modifications in DNA Repair

2.1
H2A and H2B

A modification that occurs specifically at sites of DSBs is the rapid and specific phosphorylation of histone H2AX on serine 139 in mammals (serine 129 in yeast) (Rogakou et al. 1998). This phosphorylated histone is often referred to as γ-H2AX, and for consistency will also be termed γ-H2AX in this chapter. Mammalian histone H2AX is a variant of H2A and accounts for approximately 10% of total histone H2A (Rogakou et al. 1998). In yeast the histone H2A subtypes, HTA1 and HTA2 are orthologous to the mammalian H2AX, and like H2AX, phosphorylation of yeast H2As are also implicated in the repair of DSBs (Downs et al. 2000). The kinases that phosphorylate yeast H2As are the phosphotidylinositol-3 kinase-like family members Tel1 and Mec1, which are orthologues of the ATM (Ataxia telengiectasia-mutated) and ATR (Ataxia telengiectasia-related) proteins in mammals (Burma et al. 2001; Downs et al. 2004; Paull et al. 2000; Ward and Chen 2001).

A study in budding yeast that investigated the specific localization of γ-H2AX in the chromatin region surrounding a DSB, which is induced by the HO endonuclease at the MAT locus, found that the highest amount of γ-H2AX induction occurred in the region approximately 3–5 kilobases from the DSB. However, a detectable chromatin immunoprecipitation (ChIP) γ-H2AX signal was detected within a 50-kilobase region of the DSB site. Interestingly, only low levels of γ-H2AX were detected in the immediate region surrounding the DSB (Shroff et al. 2004). This is a somewhat surprising result considering both the Mec1 and Tel1 kinases are detectable in the adjacent regions near the DSB (Kondo et al. 2001; Melo et al. 2001; Nakada et al. 2003; Rouse and Jackson 2002). The lack of detectable γ-H2AX in this area surrounding the DSB may be due to technical difficulties, such as the masking of the γ-H2AX signal by associated repair proteins, or possibly because of the specific regulation of H2AX phosphorylation that restricts the localization of γ-H2AX, such as rapid γ-H2AX turnover or possibly a γ-H2AX phosphatase.

This mutation of H2AX serine 129 in yeast has been shown to affect the checkpoint-blind repair of DSBs during S-phase (Redon et al. 2003). However, the importance of H2AX phosphorylation in DNA repair has been best demonstrated in higher eukaryotes through the development and analysis of a H2AX knockout mouse. Embryonic stem cells from this knockout mouse were shown to be more sensitive than wild-type cells to DNA damage formation by ionizing radiation (Bassing et al. 2002). Furthermore, H2AX deficiency in mice results in genomic instability and cancer predisposition, thus demonstrating that H2AX is critical for the repair of DNA lesions (Bassing et al. 2003; Celeste et al. 2002, 2003a). While investigating a specific role for γ-H2AX in DNA repair, researchers using fluorescent microscopy studies found a function for γ-H2AX that is consistent with its rapid induction in the chromatin regions around DSB sites. Specifically, it was found that γ-H2AX was important for the stable retention of DSB-induced foci that consist of proteins involved in DNA repair, such as BRCA1, 53BP1/Crb2, and NBS1 (Celeste et al. 2003b; Nakamura et al. 2004; Paull et al. 2000).

Recently published studies in yeast have characterized a relationship between γ-H2AX and chromatin-modifying complexes in DSB repair. Specifically, it was found that the recruitment of the INO80 chromatin-remodeling complex to DSBs is dependent on its association with the DNA damage-induced γ-H2AX (Downs et al. 2004; Morrison et al. 2004; van Attikum et al. 2004). Accordingly, the recruitment of INO80 to the DSB is greatly reduced in strains that lack the Mec1 (ATR) and Tel1 (ATM) kinases, as well as yeast strains expressing a mutant H2A that cannot be phosphorylated (Morrison et al. 2004; van Attikum et al. 2004). The interaction between INO80 and γ-H2AX, and also the recruitment of INO80 to the DSB, is greatly diminished

in strains that lack the Nhp10 and the Ies3 (INO eighty subunit 3) subunits
of the INO80 complex (Morrison et al. 2004). The association of the Ies3
subunit in the INO80 complex is dependent on the presence of Nhp10, an
HMG-like protein (Morrison et al. 2004; Shen et al. 2003). Therefore, these
results indicate that Nhp10, or perhaps both Nhp10 and Ies3, are responsible for establishing the interaction between INO80 and γ-H2AX at sites of
DSBs (Morrison et al. 2004). Interestingly, another recent report found that
the Arp4 subunit of both the INO80 complex and the NuA4 acetyltransferase
complex can bind to γ-H2AX peptides (Downs et al. 2004). Therefore, it is
possible that the Nhp10/Ies3 and Arp4 subunits are all involved in the interaction between these chromatin-modifying complexes and γ-H2AX.

Recently, it has also been shown in yeast that the phosphorylation of H2AX
is needed for the loading of cohesin around a 50-kilobase region surrounding
a DSB (Strom et al. 2004; Unal et al. 2004). Cohesin is a complex consisting
of Scc1, Scc3, Smc1, and Smc3 proteins that physically link sister chromatids
to each other during S-phase, which is critical for proper chromosome segregation during mitosis. However, in the past few years a DNA repair role
for cohesin has become evident. For instance, it was discovered that cohesin
formation affects post-replicative DSB repair in yeast (Sjogren and Nasmyth
2001). Additionally, in humans, cohesin subunits accumulate at DNA damage
sites and interact with the Mre11-Rad50-Nbs1 (MRN) complex, which recognizes DSBs and processes the DNA ends to create the 3′ single-strand over-
hang that is a prerequisite for homologous recombination (HR) (Kim et al.
2002). The recruitment of the cohesin complex to large chromatin regions
surrounding DSBs that contain γ-H2AX is thought to facilitate DNA repair by
maintaining sister chromatids, which serve as homologous sequence donors,
in close proximity to each other (Strom et al. 2004; Unal et al. 2004).

However, despite these research advances investigating the role of γ-H2AX
in DNA repair, mutation of serine 129 in yeast H2AX, the target of the
Mec1/Tel1 kinases, causes only modest hypersensitivity of cells to DNA-
damaging agents (Downs et al. 2000). This may be indicative of the cooperative effect of multiple histone post-translational modifications in the
process of DNA repair. For instance, serine 122 in the H2AX C-terminus
also contains a potential phosphorylation site, which when mutated results
in increased sensitivity to DNA-damaging agents and affects both HR and
non-homologous end joining (NHEJ) repair pathways. Furthermore, a strain
containing mutation of both serine 122 and 129 of H2AX results in a hypersensitive phenotype to DNA-damaging agents when compared to either single
mutant alone (Harvey et al. 2005), indicating that phosphorylation of the two
serines either contribute differently to DNA repair or that these two phosphorylation events cooperate together to affect a specific activity during DNA
repair.

Not only H2AX phosphorylation, but also H2B phosphorylation has been
observed in mammalian cells (Fernandez-Capetillo et al. 2004). Specifically,

it was found that phosphorylation of H2B on serine 14 occurs in chromatin regions surrounding ionizing radiation-induced DSBs. Following the creation of DSBs, immunofluorescent foci formation of phosphorylated serine 14 of H2B occurs after, and is dependent on, γ-H2AX induction (Fernandez-Capetillo et al. 2004). Although the kinase that phosphorylates H2B on serine 14 during DNA repair has yet to be determined, a previous role for phosphorylation of H2B serine 14 by the sterile 20 kinase (MST1) has been described during apoptosis in both higher and lower eukaryotes (Ahn et al. 2005; Cheung et al. 2003). In addition to phosphorylation, H2B is also a target for ubiquitination on lysine 123 by the Rad6 ubiquitin-conjugating enzyme and the ubiquitin ligase Bre1. It has recently been shown that lack of Rad6-Bre1 ubiquitination of H2B lysine 123 in yeast causes defects in the DNA damage checkpoint response (Giannattasio et al. 2005).

2.2
H3 and H4

Proper cell cycle checkpoint responses during DNA repair are also influenced by methylation of H3 lysine 79 by the methyltransferase Dot1, an event that is dependent on the ubiquitination of H2B lysine 123 (Giannattasio et al. 2005). However, because Dot1 affects transcription in yeast, it is not yet known if this post-translational modification of H3 by Dot1 affects the DNA damage response because of direct involvement in DNA repair or because of transcriptional regulation of DNA repair and cell cycle checkpoint genes (Singer et al. 1998).

However, it is known that methylation of lysine 79 on histone H3 is needed for the binding of Tudor domains, which are evolutionarily conserved chromodomain-like protein sequences that are present on 53BP1, a p53-binding protein that is involved in cell cycle checkpoint regulation. The orthologues of mammalian 53BP1 are Rad9 in budding yeast and Crb2 in fission yeast. Researchers investigating the involvement of methylated lysine 79 in histone H3 decreased expression of Dot1 and found that there were reduced levels of 53BP1 immunofluorescent foci following ionizing radiation treatment (Huyen et al. 2004). A similar mechanism of recruitment has been identified for Crb2 in fission yeast. The recently identified methyltransferase Set9 was found to methylate lysine 20 of histone H4, and this methylated residue is required to recruit Crb2 to sites of DNA damage (Sanders et al. 2004). Loss of either Set9 or mutation of lysine 20 of H4 results in decreased cellular viability and impaired checkpoint activation upon exposure to DNA damaging agents. However, the binding of Crb2 to DNA damage sites has also been found to be dependent on the phosphorylation of H2AX (Nakamura et al. 2004). Together, these results illustrate the notion that multiple post-translational modifications of histones may cooperate to facilitate DNA repair activities.

3
Chromatin-Modifying Complexes in DNA Repair

3.1
Histone-Modifying Complexes

Not only have histone post-translational modifications been shown to influence DNA repair, but the histone-modifying complexes themselves also affect the efficiency of DNA repair. In the past, research investigating the functions of histone-modifying complexes has focused on their role in the regulation of transcription. However, recent work has demonstrated that like transcription, DNA repair activities are also influenced by post-translational modifications of histones. For example, histone acetylation by the NuA4 acetyltransferase complex in yeast not only affects the process of transcription, but is also required for DSB repair (Bird et al. 2002; Downs et al. 2004; Nourani et al. 2001). Histone acetylation has also been shown to be involved in DNA repair in higher eukaryotes. Specifically, the activity of the human Tip60 histone acetyltransferase complex has been found to regulate the repair of DNA lesions because inhibition of Tip60 acetyltransferase activity results in the accumulation of DSBs following exposure to γ-irradiation (Ikura et al. 2000).

Histone acetylation has also been found to be involved in recruiting chromatin-remodeling complexes to sites of DSBs. For instance, it was found that loss of acetylation conferred by the NuA4 complex results in reduced association of chromatin-remodeling complexes, such as INO80 and/or SWR1, to DSBs (Downs et al. 2004). As previously noted, the binding of INO80 to DSBs is also dependent on the association of the complex with phosphorylated H2AX by the Mec1/Tel1 kinases (Morrison et al. 2004; van Attikum et al. 2004). Interestingly, it was also recently discovered that components of the NuA4 acetyltransferase complex bind to phosphorylated H2AX (Downs et al. 2004). Because NuA4 has previously been implicated in DNA repair (Bird et al. 2002), it is postulated that this chromatin-modifying complex is also recruited to sites of DNA damage through this interaction. Therefore, both the acetyltransferase activity of NuA4 and the kinase activity of Mec1/Tel1 may cooperate to facilitate binding of chromatin-remodeling complexes to DSB sites. Together, these data may demonstrate that the activity of these histone-modifying complexes in areas around DSBs occurs in a specific interdependent sequential order.

3.2
Chromatin-Remodeling Complexes

In addition to histone-modifying complexes, chromatin-remodeling complexes also assist many nuclear processes. The activity of these complexes is ATP-dependent, as they use the energy of ATP hydrolysis to alter chromatin by such mechanisms as generating DNA superhelical torsion, disrupting DNA/histone contacts, and nucleosome repositioning (Tsukiyama 2002). As discussed earlier, all the ATP-dependent chromatin-remodeling complexes are classified in the SWI/SNF chromatin-remodeling superfamily by the presence of a SNF2-like DEAD/H(SF2) ATPase subunit within the complexes (Eisen et al. 1995). As with histone-modifying complexes, the vast majority of investigations on the role of chromatin-remodeling complexes in cellular processes have been on transcription. Indeed, all four subfamilies (SWI/SNF, ISWI, CHD, and INO80) in the chromatin-remodeling complex superfamily greatly influence this process (Shen et al. 2000; Tsukiyama 2002).

However, one distant member of the SWI/SNF subfamily, Rad54, has been shown to have a specific function in DNA repair. Rad54 is a member of the *RAD52* epistasis group that interacts with and assists Rad51 during HR (Eisen et al. 1995; Mazin et al. 2003; Sugawara et al. 2003; Wolner et al. 2003). Specifically, Rad54 facilitates HR at a step following single-strand resection, but before new DNA synthesis occurs (Sugawara et al. 2003; Wolner et al. 2003). Although the importance of Rad54 in DNA repair is well-characterized, its function in chromatin remodeling has not been clearly established (Alexeev et al. 2003; Alexiadis and Kadonaga 2002; Jaskelioff et al. 2003). However, as previously mentioned, recent developments report that bona fide chromatin-remodeling complexes are involved in DNA repair, thus exposing a novel function for ATP-dependent chromatin-remodeling complexes.

Although the ATPase subunit in each chromatin-remodeling complex contains helicase motifs, the only chromatin-remodeling complex that has been shown to exhibit in vitro helicase activity is the INO80 complex (Shen et al. 2000). Unlike other subfamilies in the SWI/SNF superfamily, members of the INO80 subfamily contain two RuvB-like proteins, Rvb1 and Rvb2 (Mizuguchi et al. 2004; Shen et al. 2000). In prokaryotes, the RuvB helicase forms a complex that is involved in DNA Holliday Junction branch migration during HR (Kanemaki et al. 1997; Tsaneva et al. 1992). The identification of these helicases in the INO80 complex presented the first evidence that the complex may be involved in DNA repair and/or recombination. Accordingly, yeast strains that lack a functional Ino80 ATPase are sensitive to DNA damaging agents, such as ultraviolet light (UV), ionizing radiation (IR), and alkylating agents (MMS) (Shen et al. 2000). Therefore, the INO80 complex represents a unique subfamily of chromatin-remodeling enzymes that contains DNA repair-related proteins and is involved in DNA repair activities.

As mentioned, the histone acetyltransferase activity of the mammalian Tip60 complex is needed for efficient DNA repair (Ikura et al. 2000). Like chromatin-remodeling complexes, the Tip60 complex contains an ATPase subunit, and like the INO80 complex, Tip60 contains RuvB-like proteins (Doyon et al. 2004). Therefore, the Tip60 complex represents a unique complex that has both histone-modifying activity and chromatin-remodeling activity. Recently a report by Kusch et al. demonstrated that the *Drosophila melanogaster* homologue of Tip60 (dTip60) preferentially binds to and acetylates nucleosomal phosphorylated H2Av, which is the *Drosophila* homologue of γ-H2AX (Kusch et al. 2004). The dTip60 complex also catalyzes the exchange of phosphorylated H2Av with unmodified H2Av within chromatin (Kusch et al. 2004). Consequently, *Drosophila* cells that lack a functional dTip60 complex lose the transient acetylation of H2Av that normally occurs after exposure to γ-irradiation (Kusch et al. 2004). Additionally, phosphorylated H2Av immunofluorescent foci persists in these mutant cells following exposure to DNA-damaging agents (Kusch et al. 2004).

Additional studies have also uncovered a DNA repair role for chromatin-remodeling complexes that have previously been characterized as transcriptional regulators, such as SWI/SNF and RSC (Chai et al. 2005; Shim et al. 2005). This role for chromatin-remodeling complexes in DNA repair is believed to be evolutionarily conserved because chromatin-remodeling complexes have been found to affect DNA repair processes in both higher and lower eukaryotes (Chai et al. 2005; Downs et al. 2004; Fritsch et al. 2004; Kusch et al. 2004; Morrison et al. 2004; Shim et al. 2005; van Attikum et al. 2004). In conclusion, these studies firmly establish the role of chromatin-remodeling complexes in DNA repair and demonstrate that this process often utilizes histone modifications to recruit complexes that can manipulate chromatin in order to facilitate repair.

4
Future Directions

4.1
Additional Chromatin Modifiers in DNA Repair

Not only have the previously discussed chromatin-remodeling complexes been implicated in DNA repair process, but other chromatin-remodeling complexes may also have roles in DNA repair, such as the SWR1 complex. SWR1 is currently the only other identified member of the Rvb1/2-containing INO80 chromatin-remodeling subfamily (Mizuguchi et al. 2004). In *S. cerevisiae*, SWR1 has been found to catalyze the exchange of the histone variant H2AZ into chromatin to regulate gene expression and control the spread of heterochromatin (Mizuguchi et al. 2004). Like INO80, yeast strains lacking a functional Swr1 ATPase also display increased sensitivities to DNA damaging agents, such as MMS and UV light (Mizuguchi et al. 2004). Interestingly, the SWR1 complex also associates with γ-H2AX, although this interaction is not as robust as that of INO80 for γ-H2AX (Morrison et al. 2004). Nevertheless, these results suggest that the SWR1 complex may also be involved in the repair of DNA lesions. Furthermore, because both INO80 and SWR1 share similarities with Tip60 it can be postulated that INO80 and/or SWR1 may serve a similar function in yeast to that of Tip60 in *Drosophila*, which is the exchange of γ-H2AX with unmodified H2A during the repair of DNA. As previously mentioned, recent studies have also implicated the RSC and SWI/SNF complexes in DSB repair (Chai et al. 2005; Shim et al. 2005). Therefore, the repair of a single DSB may require the collaboration of many chromatin-modifying activities, as has been shown for the transcriptional regulation of certain genes.

4.2
Recruitment and Function of Chromatin Modifiers in DNA Repair

One potential way to direct the activity of chromatin-modifying complexes to specific sites of DNA lesions is for these complexes to recognize and bind various histone post-translational modifications that occur in response to DNA damage within the chromatin regions surrounding DNA lesions. As mentioned, an example of this method of recruitment has been demonstrated for the binding of the INO80 complex to DSB sites by association of the complex with γ-H2AX (Morrison et al. 2004; van Attikum et al. 2004). In this particular situation, the chromatin involved in the DNA damage is modified in order to recruit a complex or complexes that modulate the chromatin environment so that efficient DNA repair can occur. The binding of proteins to specifically modified histones, often referred to as a "histone code", has been previously proposed for the process of transcription (Jenuwein and Allis 2001; Strahl and Allis 2000). The recent studies that have been discussed in this chapter also suggest that a histone code may exist for DNA repair.

However, it should be noted that alternative or complementary mechanisms to a potential DNA repair histone code hypothesis might exist. For instance, some chromatin-modifying complexes may also be recruited through direct association with DNA repair proteins or damaged DNA itself. While these potential mechanisms remain to be investigated, recent reports clearly demonstrate that histone modifications are able to direct the recruitment of chromatin-modifying complexes to sites of DNA repair.

Chromatin-modifying activities might be involved in DNA repair in several ways. It is thought that chromatin modifications might affect DNA repair by providing the repair machinery with an exposed or open chromatin environment that facilitates the recruitment of DNA repair proteins. However, it can also be argued that chromatin remodeling is needed to form a compact chromatin structure, which will hold broken DNA ends close to each other. Furthermore, chromatin remodeling might also assist in the restoration of the chromatin structure after the DNA damage has been repaired.

Given the recent advances, we now know that many chromatin-modifying complexes are involved in DNA repair pathways. However, this research field is still in its infancy and it is not known precisely what function each of these chromatin-modifying complexes has during DNA repair. For example, what happens to the chromatin structure around DSBs during repair remains largely unknown, despite some indications that histone loss might be involved (Tsukuda et al. 2005). It is possible that chromatin-remodeling activities are required to "slide" nucleosomes at specific sites to allow repair machinery to bind or function. It is also possible that the histones around the DSB are being actively exchanged during repair. The turnover of γ-H2AX around the DSB can be achieved by such a histone exchange mechanism. As previously discussed, this exchange may be catalyzed by dTip60 or equivalent complexes in other organisms (Kusch et al. 2004). In yeast, the combination of several chromatin-modifying activities, such as INO80, SWR1 and NuA4 might be needed to achieve this proposed histone exchange at DSB. Alternatively, dephosphorylaton of γ-H2AX may also lead to the removal of γ-H2AX in chromatin surrounding DSBs following the completion of DNA repair. It is likely that multiple mechanisms are involved, much like the complex regulation of chromatin modifications in gene regulation, where histone modification and chromatin-remodeling events are precisely choreographed depending on the transcriptional requirements of a specific gene. Therefore, it is of interest to determine whether such potential sequences of events actually exist for the repair of a DSB.

Additionally, it has not yet been clearly demonstrated whether chromatin-remodeling complexes are involved in HR or NHEJ, or perhaps both. Synthetic genetic analyses (SGA) have demonstrated that components of the INO80 complex genetically interact with several members of the homologous recombination *RAD52* epistasis group (Morrison et al. 2004). This further supports data that demonstrate a role for INO80 in DNA repair but does not definitively establish INO80 in either HR or NHEJ. However, a recent report by Fritsch et al. implicates INO80 in HR but not NHEJ in plants (Fritsch et al. 2004). A mechanism for the involvement of INO80 in HR is supported by data presented in the publication by van Attikum et al., which shows that mutant yeast strains of the INO80 complex are defective in the single-strand DNA resection that occurs prior to strand invasion in HR (van Attikum et al. 2004). In addition, this same report also presented data demonstrating that mutant yeast strains of the INO80 complex are defective in NHEJ (van Attikum et al. 2004). Therefore, these studies suggest that INO80 is involved in both HR and NHEJ in yeast but not in plants. However, a recent report investigating the chromatin-remodeling dynamics during DNA repair in yeast found that although the INO80 complex is involved in histone eviction during DNA repair, a role for the INO80 complex in single-strand resection and NHEJ was not found (Tsukuda et al. 2005).

The activities of the RSC and SWI/SNF chromatin-remodeling complexes in yeast have also been investigated. Researchers found that subunits of the RSC complex influence the NHEJ repair pathway (Shim et al. 2005). Additionally, it was discovered that both the SWI/SNF and RSC complexes affected the HR repair pathway, although SWI/SNF appears to influence the early steps of

HR preceding strand invasion, while RSC affects a relatively late step of HR, such as post-synaptic ligation.

Despite these initial research advances, there is still much to be discovered regarding the involvement of various chromatin-modifying complexes in different organisms and whether these complexes are specialized in the repair of specific DNA lesions. For instance, the precise in vivo chromatin-remodeling activities, such as the nucleosome sliding and histone exchange that occur during repair, remain to be determined. It is tempting to speculate that the activity of these complexes may actually be stimulated by specific histone modifications.

4.3
Chromatin Modifications and Cancer

Cancer cells evolve through a multistep process that provides cells with a proliferative advantage through the attainment of several genetic alterations. Disruptions in repair pathways are one way to accelerate the accumulation of genetic alterations. These include genes such as those involved in nucleotide excision repair, mismatch repair, and double-strand break repair. These disruptions directly contribute to neoplastic growth in inheritable diseases such as xeroderma pigmentosum, hereditary non-polyposis colorectal cancer, ataxia telangiectasia, and Nijmegen breakage syndrome, as well as cancers associated with the loss of p53 and BRCA1/2 function, to name a few (Lengauer et al. 1998).

Clearly, the link between DNA repair proteins and cancer is widely acknowledged. However, the connection between chromatin modifiers and carcinogenesis has not been widely explored. A potential link does exist between certain transcriptional regulators and carcinogenesis because these transcription factors, like the tumor suppressor retinoblastoma protein, associate with chromatin modifiers in order to facilitate transcriptional regulation (Brehm et al. 1998; Magnaghi-Jaulin et al. 1998; Morrison et al. 2002; Nielsen et al. 2001). However, a few recent studies have also suggested a link between DNA repair-related chromatin modifiers and cancer. As previously discussed, there is evidence that demonstrates a requirement for H2AX phosphorylation in the maintenance of genomic integrity. Specifically, it was shown that H2AX deficiency in mice causes genomic instability and cancer predisposition (Bassing et al. 2003; Celeste et al. 2002, 2003a).

Not only histone modification, but also chromatin-remodeling complexes have been implicated in cancer. For instance, bi-allelic deletion or mutation of *SNF5*, a subunit of the SWI/SNF chromatin-remodeling complex, has been found in malignant rhabdoid tumors that are caused by an aggressive pediatric cancer (Klochendler-Yeivin et al. 2002; Neely and Workman 2002; Versteege et al. 1998). Despite these initial findings, very little is known about the specific role of chromatin remodeling in DNA repair and carcinogenesis. Since chromatin modifications affect transcription, it is possible that the chromatin modifications observed in cancer cells cause a deregulation of genes that contribute to cancer. Alternatively, because many chromatin-modifying complexes have been implicated in DNA repair, alterations in chromatin modifications may affect genome integrity as a result of defects in the DNA repair process.

4.4
Summary

The significant role of chromatin-modifying complexes in DNA repair has now become evident. Recent advances have presented an emerging model of DNA repair in which a dependent relationship between chromatin-remodeling complexes and histone modifications exists to coordinate the process of DNA damage repair. However, there is still much to be learned about the role of chromatin-modifying activities in DNA repair. The precise roles of specific chromatin-modifying activities in the various DNA repair pathways have just begun to be determined. Also, the impact of these chromatin-modifying activities in disease progression has yet to be comprehensively studied.

References

Ahn SH, Cheung WL, Hsu JY, Diaz RL, Smith MM, Allis CD (2005) Sterile 20 kinase phosphorylates histone H2B at serine 10 during hydrogen peroxide-induced apoptosis in S. cerevisiae. Cell 120:25–36

Alexeev A, Mazin A, Kowalczykowski SC (2003) Rad54 protein possesses chromatin-remodeling activity stimulated by the Rad51-ssDNA nucleoprotein filament. Nat Struct Biol 10:182–186

Alexiadis V, Kadonaga JT (2002) Strand pairing by Rad54 and Rad51 is enhanced by chromatin. Genes Dev 16:2767–2771

Armstrong JA, Emerson BM (1998) Transcription of chromatin: these are complex times. Curr Opin Genet Dev 8:165–172

Bassing CH, Chua KF, Sekiguchi J, Suh H, Whitlow SR, Fleming JC, Monroe BC, Ciccone DN, Yan C, Vlasakova K, Livingston DM, Ferguson DO, Scully R, Alt FW (2002) Increased ionizing radiation sensitivity and genomic instability in the absence of histone H2AX. Proc Natl Acad Sci USA 99:8173–8178

Bassing CH, Suh H, Ferguson DO, Chua KF, Manis J, Eckersdorff M, Gleason M, Bronson R, Lee C, Alt FW (2003) Histone H2AX: a dosage-dependent suppressor of oncogenic translocations and tumors. Cell 114:359–370

Becker PB, Horz W (2002) ATP-dependent nucleosome remodeling. Annu Rev Biochem 71:247–273

Bird AW, Yu DY, Pray-Grant MG, Qiu Q, Harmon KE, Megee PC, Grant PA, Smith MM, Christman MF (2002) Acetylation of histone H4 by Esa1 is required for DNA double-strand break repair. Nature 419:411–415

Brehm A, Miska EA, McCance DJ, Reid JL, Bannister AJ, Kouzarides T (1998) Retinoblastoma protein recruits histone deacetylase to repress transcription. Nature 391:597–601

Burma S, Chen BP, Murphy M, Kurimasa A, Chen DJ (2001) ATM phosphorylates histone H2AX in response to DNA double-strand breaks. J Biol Chem 276:42462–42467

Burns LG, Peterson CL (1997) The yeast SWI-SNF complex facilitates binding of a transcriptional activator to nucleosomal sites in vivo. Mol Cell Biol 17:4811–4819

Celeste A, Difilippantonio S, Difilippantonio MJ, Fernandez-Capetillo O, Pilch DR, Sedelnikova OA, Eckhaus M, Ried T, Bonner WM, Nussenzweig A (2003a) H2AX haploinsufficiency modifies genomic stability and tumor susceptibility. Cell 114:371–383

Celeste A, Fernandez-Capetillo O, Kruhlak MJ, Pilch DR, Staudt DW, Lee A, Bonner RF, Bonner WM, Nussenzweig A (2003b) Histone H2AX phosphorylation is dispensable for the initial recognition of DNA breaks. Nat Cell Biol 5:675–679

Celeste A, Petersen S, Romanienko PJ, Fernandez-Capetillo O, Chen HT, Sedelnikova OA, Reina-San-Martin B, Coppola V, Meffre E, Difilippantonio MJ, Redon C, Pilch DR, Olaru A, Eckhaus M, Camerini-Otero RD, Tessarollo L, Livak F, Manova K, Bonner WM, Nussenzweig MC, Nussenzweig A (2002) Genomic instability in mice lacking histone H2AX. Science 296:922–927

Chai B, Huang J, Cairns BR, Laurent BC (2005) Distinct roles for the RSC and Swi/Snf ATP-dependent chromatin remodelers in DNA double-strand break repair. Genes Dev 19:1656–1661

Cheung WL, Ajiro K, Samejima K, Kloc M, Cheung P, Mizzen CA, Beeser A, Etkin LD, Chernoff J, Earnshaw WC, Allis CD (2003) Apoptotic phosphorylation of histone H2B is mediated by mammalian sterile twenty kinase. Cell 113:507–517

Downs JA, Allard S, Jobin-Robitaille O, Javaheri A, Auger A, Bouchard N, Kron SJ, Jackson SP, Cote J (2004) Binding of chromatin-modifying activities to phosphorylated histone H2A at DNA damage sites. Mol Cell 16:979–990

Downs JA, Lowndes NF, Jackson SP (2000) A role for Saccharomyces cerevisiae histone H2A in DNA repair. Nature 408:1001–1004

Doyon Y, Selleck W, Lane WS, Tan S, Cote J (2004) Structural and functional conservation of the NuA4 histone acetyltransferase complex from yeast to humans. Mol Cell Biol 24:1884–1896

Eisen JA, Sweder KS, Hanawalt PC (1995) Evolution of the SNF2 family of proteins: subfamilies with distinct sequences and functions. Nucleic Acids Res 23:2715–2723

Fernandez-Capetillo O, Allis CD, Nussenzweig A (2004) Phosphorylation of histone H2B at DNA double-strand breaks. J Exp Med 199:1671–1677

Fritsch O, Benvenuto G, Bowler C, Molinier J, Hohn B (2004) The INO80 protein controls homologous recombination in Arabidopsis thaliana. Mol Cell 16:479–485

Fyodorov DV, Kadonaga JT (2001) The many faces of chromatin remodeling: SWItching beyond transcription. Cell 106:523–525

Giannattasio M, Lazzaro F, Plevani P, Muzi-Falconi M (2005) The DNA damage checkpoint response requires histone H2B ubiquitination by Rad6-Bre1 and H3 methylation by Dot1. J Biol Chem 280:9879–9886

Harvey AC, Jackson SP, Downs JA (2005) Saccharomyces cerevisiae histone H2A Ser122 facilitates DNA repair. Genetics 170:543–553

Hirschhorn JN, Brown SA, Clark CD, Winston F (1992) Evidence that SNF2/SWI2 and SNF5 activate transcription in yeast by altering chromatin structure. Genes Dev 6:2288–2298

Huyen Y, Zgheib O, Ditullio RA Jr, Gorgoulis VG, Zacharatos P, Petty TJ, Sheston EA, Mellert HS, Stavridi ES, Halazonetis TD (2004) Methylated lysine 79 of histone H3 targets 53BP1 to DNA double-strand breaks. Nature 432:406–411

Ikura T, Ogryzko VV, Grigoriev M, Groisman R, Wang J, Horikoshi M, Scully R, Qin J, Nakatani Y (2000) Involvement of the TIP60 histone acetylase complex in DNA repair and apoptosis. Cell 102:463–473

Jaskelioff M, Van Komen S, Krebs JE, Sung P, Peterson CL (2003) Rad54p is a chromatin remodeling enzyme required for heteroduplex DNA joint formation with chromatin. J Biol Chem 278:9212–9218

Jenuwein T, Allis CD (2001) Translating the histone code. Science 293:1074–1080

Kadonaga JT (1998) Eukaryotic transcription: an interlaced network of transcription factors and chromatin-modifying machines. Cell 92:307–313

Kanemaki M, Makino Y, Yoshida T, Kishimoto T, Koga A, Yamamoto K, Yamamoto M, Moncollin V, Egly JM, Muramatsu M, Tamura T (1997) Molecular cloning of a rat 49-kDa TBP-interacting protein (TIP49) that is highly homologous to the bacterial RuvB. Biochem Biophys Res Commun 235:64–68

Kim JS, Krasieva TB, LaMorte V, Taylor AM, Yokomori K (2002) Specific recruitment of human cohesin to laser-induced DNA damage. J Biol Chem 277:45149–45153

Kingston RE, Narlikar GJ (1999) ATP-dependent remodeling and acetylation as regulators of chromatin fluidity. Genes Dev 13:2339–2352

Klochendler-Yeivin A, Muchardt C, Yaniv M (2002) SWI/SNF chromatin remodeling and cancer. Curr Opin Genet Dev 12:73–79

Kondo T, Wakayama T, Naiki T, Matsumoto K, Sugimoto K (2001) Recruitment of Mec1 and Ddc1 checkpoint proteins to double-strand breaks through distinct mechanisms. Science 294:867–870

Kornberg RD, Lorch Y (1999) Twenty-five years of the nucleosome, fundamental particle of the eukaryote chromosome. Cell 98:285–294

Kusch T, Florens L, Macdonald WH, Swanson SK, Glaser RL, Yates JR 3rd, Abmayr SM, Washburn MP, Workman JL (2004) Acetylation by Tip60 is required for selective histone variant exchange at DNA lesions. Science 306:2084–2087

Lengauer C, Kinzler KW, Vogelstein B (1998) Genetic instabilities in human cancers. Nature 396:643–649

Luger K, Richmond TJ (1998) DNA binding within the nucleosome core. Curr Opin Struct Biol 8:33–40

Magnaghi-Jaulin L, Groisman R, Naguibneva I, Robin P, Lorain S, Le Villain JP, Troalen F, Trouche D, Harel-Bellan A (1998) Retinoblastoma protein represses transcription by recruiting a histone deacetylase. Nature 391:601–605

Mazin AV, Alexeev AA, Kowalczykowski SC (2003) A novel function of Rad54 protein. Stabilization of the Rad51 nucleoprotein filament. J Biol Chem 278:14029–14036

Melo JA, Cohen J, Toczyski DP (2001) Two checkpoint complexes are independently recruited to sites of DNA damage in vivo. Genes Dev 15:2809–2821

Mizuguchi G, Shen X, Landry J, Wu WH, Sen S, Wu C (2004) ATP-driven exchange of histone H2AZ variant catalyzed by SWR1 chromatin remodeling complex. Science 303:343–348

Morrison AJ, Highland J, Krogan NJ, Arbel-Eden A, Greenblatt JF, Haber JE, Shen X (2004) INO80 and γ-H2AX interaction links ATP-dependent chromatin remodeling to DNA damage repair. Cell 119:767–775

Morrison AJ, Sardet C, Herrera RE (2002) Retinoblastoma protein transcriptional repression through histone deacetylation of a single nucleosome. Mol Cell Biol 22:856–865

Nakada D, Matsumoto K, Sugimoto K (2003) ATM-related Tel1 associates with double-strand breaks through an Xrs2-dependent mechanism. Genes Dev 17:1957–1962

Nakamura TM, Du LL, Redon C, Russell P (2004) Histone H2A phosphorylation controls Crb2 recruitment at DNA breaks, maintains checkpoint arrest, and influences DNA repair in fission yeast. Mol Cell Biol 24:6215–6230

Neely KE, Workman JL (2002) The complexity of chromatin remodeling and its links to cancer. Biochim Biophys Acta 1603:19–29

Nielsen SJ, Schneider R, Bauer UM, Bannister AJ, Morrison A, O'Carroll D, Firestein R, Cleary M, Jenuwein T, Herrera RE, Kouzarides T (2001) Rb targets histone H3 methylation and HP1 to promoters. Nature 412:561–565

Nourani A, Doyon Y, Utley RT, Allard S, Lane WS, Cote J (2001) Role of an ING1 growth regulator in transcriptional activation and targeted histone acetylation by the NuA4 complex. Mol Cell Biol 21:7629–7640

Paull TT, Rogakou EP, Yamazaki V, Kirchgessner CU, Gellert M, Bonner WM (2000) A critical role for histone H2AX in recruitment of repair factors to nuclear foci after DNA damage. Curr Biol 10:886–895

Redon C, Pilch DR, Rogakou EP, Orr AH, Lowndes NF, Bonner WM (2003) Yeast histone 2A serine 129 is essential for the efficient repair of checkpoint-blind DNA damage. EMBO Rep 4:678–684

Rogakou EP, Pilch DR, Orr AH, Ivanova VS, Bonner WM (1998) DNA double-stranded breaks induce histone H2AX phosphorylation on serine 139. J Biol Chem 273:5858–5868

Roth SY, Denu JM, Allis CD (2001) Histone acetyltransferases. Annu Rev Biochem 70:81–120

Rouse J, Jackson SP (2002) Lcd1p recruits Mec1p to DNA lesions in vitro and in vivo. Mol Cell 9:857–869

Sanders SL, Portoso M, Mata J, Bahler J, Allshire RC, Kouzarides T (2004) Methylation of histone H4 lysine 20 controls recruitment of Crb2 to sites of DNA damage. Cell 119:603–614

Shen X, Mizuguchi G, Hamiche A, Wu C (2000) A chromatin remodelling complex involved in transcription and DNA processing. Nature 406:541–544

Shen X, Ranallo R, Choi E, Wu C (2003) Involvement of actin-related proteins in ATP-dependent chromatin remodeling. Mol Cell 12:147–155

Shim EY, Ma JL, Oum JH, Yanez Y, Lee SE (2005) The yeast chromatin remodeler RSC complex facilitates end joining repair of DNA double-strand breaks. Mol Cell Biol 25:3934–3944

Shroff R, Arbel-Eden A, Pilch D, Ira G, Bonner WM, Petrini JH, Haber JE, Lichten M (2004) Distribution and dynamics of chromatin modification induced by a defined DNA double-strand break. Curr Biol 14:1703–1711

Singer MS, Kahana A, Wolf AJ, Meisinger LL, Peterson SE, Goggin C, Mahowald M, Gottschling DE (1998) Identification of high-copy disruptors of telomeric silencing in Saccharomyces cerevisiae. Genetics 150:613–632

Sjogren C, Nasmyth K (2001) Sister chromatid cohesion is required for postreplicative double-strand break repair in Saccharomyces cerevisiae. Curr Biol 11:991–995

Strahl BD, Allis CD (2000) The language of covalent histone modifications. Nature 403:41–45

Strom L, Lindroos HB, Shirahige K, Sjogren C (2004) Postreplicative recruitment of cohesin to double-strand breaks is required for DNA repair. Mol Cell 16:1003–1015

Sudarsanam P, Iyer VR, Brown PO, Winston F (2000) Whole-genome expression analysis of snf/swi mutants of Saccharomyces cerevisiae. Proc Natl Acad Sci USA 97:3364–3369

Sugawara N, Wang X, Haber JE (2003) In vivo roles of Rad52, Rad54, and Rad55 proteins in Rad51-mediated recombination. Mol Cell 12:209–219

Tsaneva IR, Muller B, West SC (1992) ATP-dependent branch migration of Holliday junctions promoted by the RuvA and RuvB proteins of E. coli. Cell 69:1171–1180

Tse C, Fletcher TM, Hansen JC (1998a) Enhanced transcription factor access to arrays of histone H3/H4 tetramer. DNA complexes in vitro: implications for replication and transcription. Proc Natl Acad Sci USA 95:12169–12173

Tse C, Sera T, Wolffe AP, Hansen JC (1998b) Disruption of higher-order folding by core histone acetylation dramatically enhances transcription of nucleosomal arrays by RNA polymerase III. Mol Cell Biol 18:4629–4638

Tsukiyama T (2002) The in vivo functions of ATP-dependent chromatin-remodelling factors. Nat Rev Mol Cell Biol 3:422–429

Tsukiyama T, Wu C (1995) Purification and properties of an ATP-dependent nucleosome remodeling factor. Cell 83:1011–1020

Tsukuda T, Fleming AB, Nickoloff JA, Osley MA (2005) Chromatin remodelling at a DNA double-strand break site in Saccharomyces cerevisiae. Nature 438:379–383

Unal E, Arbel-Eden A, Sattler U, Shroff R, Lichten M, Haber JE, Koshland D (2004) DNA damage response pathway uses histone modification to assemble a double-strand break-specific cohesin domain. Mol Cell 16:991–1002

van Attikum H, Fritsch O, Hohn B, Gasser SM (2004) Recruitment of the INO80 complex by H2A phosphorylation links ATP-dependent chromatin remodeling with DNA double-strand break repair. Cell 119:777–788

Versteege I, Sevenet N, Lange J, Rousseau-Merck MF, Ambros P, Handgretinger R, Aurias A, Delattre O (1998) Truncating mutations of hSNF5/INI1 in aggressive paediatric cancer. Nature 394:203–206

Ward IM, Chen J (2001) Histone H2AX is phosphorylated in an ATR-dependent manner in response to replicational stress. J Biol Chem 276:47759–47762

Wolner B, van Komen S, Sung P, Peterson CL (2003) Recruitment of the recombinational repair machinery to a DNA double-strand break in yeast. Mol Cell 12:221–232

Workman JL, Kingston RE (1998) Alteration of nucleosome structure as a mechanism of transcriptional regulation. AnnRev Biochem 67:545–579

Results Probl Cell Differ (41)
L. Brehon: Chromatin Dynamics in Cellular Function
DOI 10.1007/005/Published online: 30 March 2006
© Springer-Verlag Berlin Heidelberg 2006

Mechanisms for Nucleosome Movement by ATP-dependent Chromatin Remodeling Complexes

Anjanabha Saha · Jacqueline Wittmeyer · Bradley R. Cairns (✉)

Department of Oncological Sciences and Howard Hughes Medical Institute,
Huntsman Cancer Institute, University of Utah School of Medicine, 2000 Circle of Hope,
Salt Lake City, UT 84112, USA
brad.cairns@hci.utah.edu

Abstract Chromatin remodeling complexes (remodelers) are a set of diverse multi-protein machines that reposition and restructure nucleosomes. Remodelers are specialized, containing unique proteins that assist in targeting, interaction with modified nucleosomes, and performing specific chromatin tasks. However, all remodelers contain an ATPase domain that is highly similar to known DNA translocases/helicases, suggesting that DNA translocation is a property common to all remodelers. Here we examine the different reactions they perform in vitro, focusing on the SWI/SNF and the ISWI complexes, and explore how DNA translocation might be utilized to execute various remodeling processes.

1
Introduction

Nucleosomes are active participants in all chromosomal processes including transcription, DNA repair, replication, and the specialized function of centromeres and telomeres (Kornberg and Lorch 1999; Wu and Grunstein 2000). An extensive and evolving literature supports the roles of histone modifications and chromatin structural changes in guiding these processes. Histone modifications provide marks that recruit and regulate factors, whereas structural changes such as nucleosome repositioning or ejection help to provide regulated access of factors to the underlying DNA (Almer et al. 1986; Jenuwein and Allis 2001; Narlikar et al. 2002). Interestingly, histone modifications themselves have little impact on intrinsic nucleosome mobility or stability. Rather, nucleosomes are mobilized by the action of remodelers (Owen-Hughes 2003). Together, nucleosome modifications and structural changes help to order factor recruitment and DNA accessibility. A clear example of this coordinated regulation is in gene transcription, where different histone modifiers and remodelers build chromatin of different instructive character at the promoter, transcription initiation site, ORF, and terminator (Narlikar et al. 2002). Here, different modifications in each region guide transcription factors to their correct location on the gene, whereas remodeling factors ensure that nucleosomes function as mobile and active regulatory participants, rather than simply as obstacles.

As chromatin is specialized by modification of nucleosomes in particular regions of genes, the complexes that mobilize, eject, or reconstruct nucleosomes may have co-evolved to perform specialized tasks at these locations. However, remodeler specialization extends beyond gene transcription to include processes such as chromatin assembly and DNA repair (Cairns 2005).

In keeping with their specialization in vivo, each remodeler displays unique properties in vitro as different products are observed following their action on mononucleosomes or chromatin arrays. These observations might suggest that each remodeler imposes a different mechanism for nucleosome restructuring (Fan et al. 2003). However, all remodelers contain an ATPase domain that is highly similar to that of known DNA translocases and require ATP hydrolysis for their remodeling functions. Importantly, DNA translocation has been demonstrated by several remodelers (Jaskelioff et al. 2003; Saha et al. 2002; Whitehouse et al. 2003). Furthermore, recent evidence suggests that the ATPase domains of remodelers may engage nucleosomal DNA at a similar location and remodel nucleosomes in a manner consistent with DNA translocation (Saha et al. 2005). These results raise the intriguing possibility that all remodelers utilize DNA translocation as an aspect of their mechanism, but apply and regulate this property in different ways to achieve various outcomes. Here, we will begin with an introduction to the dynamic properties of chromatin, then examine the different reactions performed by remodelers in vitro, and finally explore how DNA translocation might be utilized to execute specific remodeling tasks.

2
Nucleosome Specialization

The nucleosome is the basic repeating unit of chromatin that consists of 146 bp of DNA wrapped around a cylindrical octamer of histone proteins (Luger et al. 1997). Typically the histone octamer is constructed of the four canonical proteins, H2A, H2B, H3, and H4, although histone variants are also utilized for further specialization (Henikoff et al. 2004). Nucleosomes are dynamic in both their covalent modification state and their translational position on the DNA (Fig. 1). Covalent modifications of histone proteins (i.e. acetylation, methylation, and phosphorylation) are performed by chromatin-modifying complexes (Jenuwein and Allis 2001). These modifications added to the histones assist in the recruitment of various factors to particular loci. Here, both the type of modification on the histone proteins as well as the particular residue modified are important determinants for recruitment specificity. For example, methylation of lysine 4 on the histone H3 attracts factors involved in transcriptional activation, whereas methylation of lysine 9 on histone H3 attracts repressive factors. Thus, covalent modifications affect the state of activity of the gene or that region of chromatin (Jenuwein and Al-

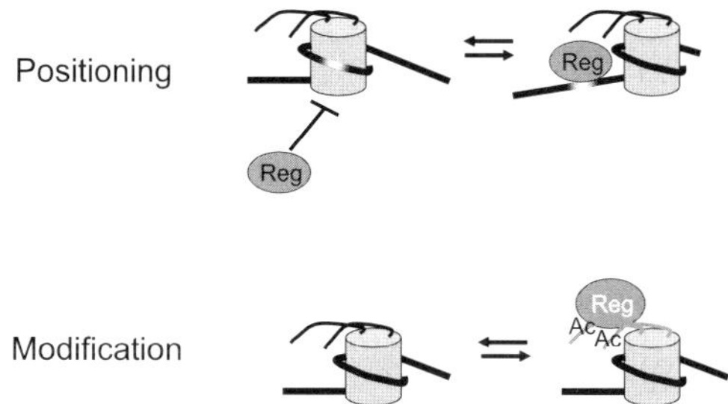

Fig. 1 Dynamic properties of the nucleosome. *Top*: nucleosome repositioning allows binding of regulatory factors (Reg) to nucleosomal DNA (*shaded white*). *Bottom*: nucleosome modification (Ac) allows binding of regulatory factors to the histone tails of the nucleosome

lis 2001). Alternatively, nucleosomes are mobilized by remodelers to assume their correct positions on DNA, which can either facilitate or impede DNA-templated processes (Owen-Hughes 2003). Many studies suggest that these specialized properties of nucleosomes work in concert to control chromatin architectural transitions.

3
The Nucleosome: A Biophysical Challenge for Remodelers

To understand the biological and mechanistic significance of remodelers, one must appreciate the obstacle that the nucleosome presents. A key biophysical feature is that the DNA contacts the octamer surface 14 times, that together contribute to the remarkable stability of the nucleosome, with the free energy of about 12–14 kcal/mol between the histone–DNA contacts (Gottesfeld and Luger 2001). The DNA contacts the octamer non-specifically through the sugar phosphate backbone, thus allowing different DNA sequences in the genome to be accommodated within a nucleosome (Widom 2001). Under physiological salt concentration, the octamer in the absence of DNA dissociates into its subcomponents, suggesting that histone–DNA contacts are crucial to maintain the oligomeric state of the nucleosome.

The distance between each histone–DNA contact is, on average, 10.2 bp. However, one particularly interesting feature of nucleosomes was revealed through crystal structures; the presence of localized regions where overtwisting of DNA is observed (Luger et al. 1997; Richmond and Davey 2003; Suto et al. 2003). In these regions, the DNA between two histone–DNA contacts is stretched by about 1 bp, relative to the rest of the nucleosomal DNA, such that there are only 9.5 bp within the helical turn. This results in a twist defect in that region, as the DNA is overtwisted relative to the rest of the nucleosomal DNA. DNA segments with twist defect have been observed in multiple nucleosome structures, and at different locations, suggesting that nucleosomes can accommodate structural changes in the DNA. Thus, our current conception of the nucleosome involves an octameric protein disk around which DNA is wrapped; the DNA resembles a large loaded spring bearing tension in the form of both twist and writhe, constrained by the 14 slightly flexible histone–DNA contacts.

For proteins to gain access to the nucleosomal DNA, the DNA can be translationally repositioned relative to the octamer to move sites into the linker or, alternatively, a stretch of DNA can be unwrapped from the surface of the octamer (Widom 1998). Biophysical studies have provided clear evidence that nucleosomes exist in a dynamic equilibrium between a fully wrapped state and a series of partially unwrapped states (Li and Widom 2004). The transient unwrapped state initiates from the edge of the nucleosome and can progressively move towards the dyad. Transient unwrapping could allow binding of regulatory proteins to the nucleosomal DNA and drive the equilibrium towards the unwrapped state. However, this accessibility of nucleosomal DNA is rapid, since nucleosomes remain fully wrapped for 250 ms before spontaneously unwrapping and rewrapping within 10–50 ms (Li et al. 2005). Even though spontaneous exposure allows accessibility near the edge, it does not allow efficient exposure of sites near the nucleosomal dyad where the equilibrium constant for site exposure is 10^{-4}–10^{-5} M, compared to 1–4×10^{-2} M at the edge of the nucleosome (Widom 1998).

Translational repositioning provides an additional mechanism for access but, in contrast to the 'peeling' mechanism, requires the breakage and reformation of all histone–DNA contacts. Although nucleosomes are capable of translational movement in vitro by thermal diffusion, they display slow kinetics and can be trapped in thermodynamically favored positions (Flaus and Owen-Hughes 2003b). A significant breakthrough was the discovery of chromatin remodeling complexes, and the demonstration that these complexes utilize the energy of ATP hydrolysis to promote rapid nucleosome repositioning and access to nucleosomal DNA (Owen-Hughes 2003). However, these studies also revealed that different remodelers are not identical in their treatment or selection of nucleosomes, raising interesting questions regarding diversity and specialization. Next, we explore the diversity of remodelers before considering the mechanism of nucleosome movement.

3.1
Remodeler Families: Discovery, Functions, and Properties

Remodelers consist of at least five different classes of complexes defined by their composition, in vitro activities, and in vivo functions. However, they all share a related catalytic subunit that contains a highly conserved ATPase domain.

The first remodeler identified was the yeast SWI/SNF complex, and several members of this complex were obtained through genetic screens for genes involved in gene activation (Winston and Carlson 1992). Subsequent biochemical purification and characterization led to the identification of the 11-subunit SWI/SNF complex (Cairns et al. 1994; Peterson et al. 1994). Based on sequence similarity to the yeast SWI/SNF complex, the 15-subunit yeast RSC (*remodels the structure of chromatin*) complex was identified, and shown to be essential for viability (Cairns et al. 1996). Correspondingly, related protein complexes were identified in human cells (termed hSWI/SNF or BAF/PBAF complexes), as well as in *Drosophila*, which display properties similar to their yeast counterparts (Becker and Horz 2002). SWI/SNF-related complexes are generally, but not solely, associated with transcriptional activation, and have additional roles in transcriptional elongation, DNA repair, cohesion loading, and chromosome stability (Chai et al. 2005; Corey et al. 2003; Huang et al. 2004; Martens and Winston 2003).

The ISWI (*imitation switch*) family of remodelers were originally identified in *Drosophila*, and include ACF (*ATP-utilizing chromatin assembly and remodeling factor*), NURF (*nucleosome remodeling factor*), and CHRAC (*chromatin accessibility complex*) (Ito et al. 1997; Tsukiyama and Wu 1995; Varga-Weisz et al. 1997). Notably, all three of these complexes possess the same ATPase protein, ISWI, but contain different associated subunits, which are likely important for specialized ISWI remodeler functions. Outside the ATPase domain, the ISWI protein diverges in homology from the SWI/SNF family. ISWI complexes were subsequently identified in yeast, *Xenopus*, and humans (Becker and Horz 2002). ISWI remodelers have diverse roles, but are most clearly connected with chromatin assembly, and in yeast appear to have intriguing connections to transcription elongation (Corona and Tamkun 2004; Mellor and Morillon 2004).

The Mi-2/NURD (*nucleosome remodeling and deacetylation*) family of remodelers are distinguished by the presence of a chromo (*chromatin organization modifier*) domain and a methylated DNA binding domain. Mi-2/NURD remodelers were initially identified in human cells and were shown to possess both chromatin remodeling and histone deacetylase activity (Becker and Horz 2002; Bowen et al. 2004). Chromodomain-containing remodelers have since been identified in *S. cerevisiae*, *S. pombe*, *Drosophila*, and *Xenopus*. Their compositions and functions are diverse, and currently less well understood than other remodelers (Bowen et al. 2004).

The INO80 family of remodelers were initially identified in *S. cerevisiae* and have been subsequently identified in *Arabidopsis* and humans (Shen et al. 2000). INO80 complexes also contain two AAA+ ATPase subunits (Rbv1/2) and several actin-related proteins, among many other subunits. INO80 remodelers display dual functions in both transcriptional activation and DNA repair, and may be guided to particular loci by chromatin modifications (van Attikum and Gasser 2005).

The SWR1 family is the newest subfamily of remodelers and has been identified in *S. cerevisiae*, *Drosophila*, and humans (Korber and Horz 2004).

They are involved in ATP-dependent replacement of histone H2A/H2B dimers with variant H2A.Z/H2B dimers, and represent a novel function for an ATP-dependent remodeler.

All remodelers contain a highly conserved ATPase region that is involved in energy-dependent alteration of chromatin structure, suggesting a possible common underlying mechanism. Here, we compare and contrast the properties of the two most intensively studied remodeler families: SWI/SNF and ISWI. We later follow this comparison with mechanistic models that might explain their similarities and differences.

3.2
Remodelers Elicit DNA- and/or Nucleosome-dependent ATPase Activity

Insights into functions and mechanisms of the SWI/SNF and the ISWI families of remodelers have been revealed by examining the different manner in which they select and bind substrates for activity. SWI/SNF and ISWI remodelers show little selectivity for DNA substrates, as they bind both single- and double-stranded DNA molecules greater than 20 nt/bp without regard to sequence specificity (Cairns et al. 1996; Lorch et al. 1998; Whitehouse et al. 2003). However, ISWI remodelers bind DNA with lower affinity than the SWI/SNF remodelers. Interestingly, certain ISWI remodelers display nucleotide-dependent DNA binding, a feature not observed with SWI/SNF remodelers (Fitzgerald et al. 2004).

For SWI/SNF remodelers, ATP hydrolysis is stimulated by both single- and double-stranded DNA to a similar extent (Cairns et al. 1994, 1996; Cote et al. 1994). Although nucleosomes likewise stimulate ATPase activity, they are not more effective than DNA alone. In contrast, whereas ISWI remodelers display only a modest stimulation of ATPase activity in the presence of single- and double-stranded DNA, they additionally require elements of nucleosomes, as described below (Whitehouse et al. 2003). Under optimal conditions, the DNA-stimulated ATPase activity of SWI/SNF remodelers is two- to three-fold higher than that of ISWI remodelers (Saha et al. 2002; Whitehouse et al. 2003). Additionally, SWI/SNF remodelers bind nucleosomes with three-fold higher affinity in the presence of hydrolyzable ATP than in its absence (Lorch et al. 1998), suggesting that an ATP-dependent conformational change in the remodeler alters the mode of binding.

SWI/SNF and ISWI remodelers display significant differences with respect to requirements for linker DNA emitting from the nucleosome and for the presence of histone tails. For SWI/SNF remodelers, the presence of linker DNA has no influence on nucleosome binding and ATPase activity (Saha et al. 2005). Further, neither the binding nor the activity of SWI/SNF remodelers is greatly influenced by the presence of histone tails (Guyon et al. 1999). In contrast, ISWI remodelers bind nucleosomes with linker DNA much more efficiently than those lacking a linker, and their nucleosome-dependent ATPase activity increases with the presence of linker DNA (Brehm et al. 2000). Moreover, the N-terminal tail of histone H4, in particular residues 16–19, plays a critical role for ISWI binding and greatly stimulates ISWI ATPase activity (Clapier et al. 2001). Indeed, this epitope is likely a major regulator of the biology of ISWI remodelers; substrates for ISWI remodelers may be limited to nucleosomes lacking H4K16 acetylation, a modification correlated with transcriptional activation (Corona et al. 2002). In contrast, this epitope has no demonstrated influence on SWI/SNF remodeler activity.

3.3
Nucleosome Sliding and Accessibility

SWI/SNF and ISWI remodelers share the ability to catalyze the redistribution of nucleosomes along the DNA in *cis*, but generate remarkably different distribution patterns, which likely underlie their specialization in vivo. Both SWI/SNF and ISWI complexes efficiently reposition a nucleosome along the same template without complete dissociation/reassociation of the histone octamer (Langst et al. 1999; Whitehouse et al. 1999). This suggests that nucleosome sliding is a common property of both remodelers. However, each remodeler generates a different distribution of sliding products. For example, ISWI complexes such as ACF and CHRAC have the remarkable ability to translationally phase nucleosome arrays, promoting the equal spacing of DNA between each nucleosome on the template (Ito et al. 1997; Varga-Weisz et al. 1997). In clear contrast, SWI/SNF remodelers will randomize nucleosome positioning arrays that were initially spaced.

Different sliding products are also observed with mononucleosomes (Fig. 2). SWI/SNF largely disorders the population of mononucleosomes with respect to translational positions on the octamer. However, one prominent product contains DNA recessed about 50 bp inside the nucleosome, resulting

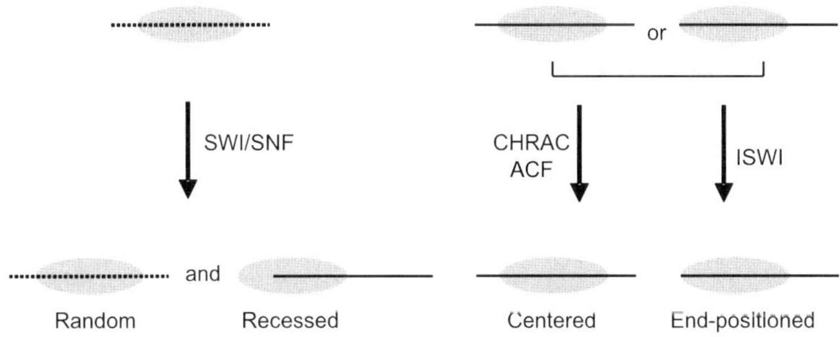

Fig. 2 Sliding properties of SWI/SNF and ISWI remodelers. Nucleosomes are depicted in 2D with the histone octamer position depicted by the *grey oval*. *Black line* indicates the position of DNA, with *dotted line* referring to DNA at random positions along the octamer

in a nucleosome species that lacks about five histone–DNA contacts (Flaus and Owen-Hughes 2003b; Kassabov et al. 2003). This product is not observed with ISWI remodelers, which display an alternative product distribution; rather than disordering the translational positions, ISWI remodelers move the population to a largely uniform translational position. Interestingly, proteins associated with ISWI help select the favored translational position. For example, the isolated ISWI protein preferentially slides an octamer positioned near the center of a DNA fragment towards the end, whereas the CHRAC or ACF complexes move octamers positioned near the DNA end to a more central position (Langst et al. 1999). Furthermore, the ISWI protein as a part of the NURF complex slides mononucleosomes towards the thermodynamically preferred position on DNA (Hamiche et al. 1999). Thus, ISWI promotes sliding, but the associated proteins influence the outcome of the sliding reaction.

Analogous to the differences in the products of sliding reactions, the SWI/SNF and ISWI remodelers display differences in their ability to provide accessibility to nucleosomal DNA. SWI/SNF remodelers can alter nucleosome structure by repositioning the DNA around the histone octamer, as demonstrated by their ability to alter the DNaseI digestion pattern of mononucleosomes (Cairns et al. 1996; Cote et al. 1994; Imbalzano et al. 1996; Kwon et al. 1994). Furthermore, SWI/SNF remodelers render mononucleosomes accessible to transcription factors, and allow restriction endonuclease (RE) accessibility in an ATP-dependent manner (Cote et al. 1994; Lorch et al. 1998; Schnitzler et al. 1998). Like SWI/SNF remodelers, ISWI remodelers also alter the DNaseI digestion pattern of nucleosomes, but lack the ability to increase RE accessibility and transcription factor binding to mononucleosomes with no or short DNA linkers (Langst et al. 1999; Tsukiyama and Wu 1995).

Taken together, nucleosome mobilization is a common property of the SWI/SNF and the ISWI remodelers, but differences in accessibility and translational products generated might suggest that each family utilizes a different mechanism for movement (Fan et al. 2003). Alternatively, and more likely, these remodelers may share a similar underlying mechanism which could be applied and regulated differently to generate alternative remodeled products; attributes that serve to specialize the remodeler for particular tasks in vivo. As remodelers share the general property of altering the position of DNA relative to the histone octamer, we consider ATP-dependent DNA translocation as a possible unifying property.

3.4
The SWI/SNF and ISWI Remodelers
are ATP-dependent Directional DNA Translocases

Over the past few years, a series of studies with the remodelers RSC, ySWI/SNF, ISWI, and Rad54 have established that remodeler ATPases are DNA translocases (Alexeev et al. 2003; Jaskelioff et al. 2003; Saha et al. 2002;

Whitehouse et al. 2003). We will first present the biochemical evidence for DNA translocation and subsequently discuss how this property can be applied to explain many of the observed nucleosomal and DNA products.

Biochemical studies with the remodeler RSC and its isolated catalytic subunit, Sth1, provided several lines of evidence for the coupling of ATPase activity to DNA translocation (Saha et al. 2002). RSC/Sth1 ATPase activity ($V_{\max}$) is proportional to DNA length, whereas K_m and K_d are largely independent of length (for DNA lengths greater than 20 bp, the minimal length required for binding). This observation is consistent with DNA translocation; if ATP hydrolysis is proportional to the distance translocated, and translocation is not rate limiting in the reaction cycle, then short DNA fragments should elicit less ATPase activity than longer fragments. This length dependence of the ATPase activity has subsequently been observed with other remodelers such as ISWI, SWI/SNF, and Rad54 (Jaskelioff et al. 2003; Whitehouse et al. 2003). These studies also suggest a processivity of about 80 bp for RSC and about 40 bp for ISWI. Further, the ATPase activity is equally effective with both single- and double-stranded DNA, suggesting that translocation likely involves tracking of the enzyme along one strand of the DNA duplex (Cairns et al. 1996; Saha et al. 2002; Whitehouse et al. 2003).

More direct evidence for DNA translocation was revealed by the capacity of remodelers to displace the third strand from a DNA triple helix in an ATP-dependent manner (Jaskelioff et al. 2003; Saha et al. 2002; Whitehouse et al. 2003). The displacement of the third strand occurs on a nicked DNA substrate as well, suggesting that displacement occurs via invasion of the triple helix by tracking of the remodeler and not simply by twisting of the DNA. Importantly, triplex displacement activity can be attributed solely to the catalytic ATPase subunit, as the catalytic subunit in isolation is effective at displacement. Interestingly, the triple helix displacement activity of both RSC and ISWI displays a $3'$ to $5'$ strand specificity, suggesting that remodelers are directional DNA translocases (Saha et al. 2005; Whitehouse et al. 2003). Together, these studies establish that remodelers couple the energy of ATP hydrolysis to unidirectional DNA translocation along the backbone of one strand of the DNA duplex.

4
Remodelers Resemble DNA Helicases/Translocases

Consistent with the property of DNA translocation, the catalytic subunit of all remodelers contains a DEAD/H-box ATPase domain belonging to the SF2 superfamily of helicases (Eisen et al. 1995; Laurent et al. 1992). This highly conserved ATPase domain that includes several DNA and RNA helicases that are demonstrated translocases (Singleton and Wigley 2002). The SF2 family also includes proteins like type I restriction enzymes, which track along double-stranded DNA but lack the strand-separation activity displayed by helicases (Murray 2000). Another SF2 family member is Rad54, a DNA repair protein that shares with remodelers the capacity to alter nucleosome positioning in vitro (Alexeev et al. 2003; Durr et al. 2005; Jaskelioff et al. 2003). Biochemical and structural studies with helicases in the SF2 and highly related SF1 families have shown that the structure and function of the catalytic regions are quite similar (Kim et al. 1998; Singleton et al. 2001; Velankar et al. 1999). They are composed of a DNA duplex destabilizing domain, coupled to a translocating motor (Singleton and Wigley 2002). Further, the primary function of the helicase domain is to act as a molecular motor, moving the enzyme along the DNA template. Strand separation is a secondary function that is coupled to, but not required for, translocation (Singleton and Wigley 2002). In fact, in PcrA translocation and strand-separation activities can be uncoupled, as certain mutations have no effect on the translocation properties, but abolish the helicase activity (Soultanas et al. 2000). Although many SF2 family members are helicases, chromatin remodelers lack strand-separation activity. This lack of helicase activity is not surprising as chromatin remodeling does not involve the creation of single-stranded regions (Cote et al. 1998).

Taken together, all studied SF2 and SF1 family members are ATP-dependent translocating enzymes. In addition, recently obtained crystal structures of Rad54 alone and in complex with DNA show a high degree of similarity in the DNA translocation domain with other known helicases, again consistent with remodelers translocating on DNA (Durr et al. 2005; Thoma et al. 2005).

4.1
DNA Translocation from an Internal Nucleosomal Site

These studies raised two key questions: how and where is translocation applied on the nucleosome? Recent advances have been made in understanding how the remodeler and the ATPase subunit engage the nucleosome. First, gel mobility shift assays with both RSC and ISWI suggest that one remodeler complex binds to a single nucleosome (Langst and Becker 2001; Lorch et al. 1998). For the ISWI family of remodelers, an elegant combination of cross-linking and high-resolution DNA footprinting revealed that ISWI binds to the nucleosome at two separate locations: an internal site near the nucleosomal dyad, and an external site involving one of the DNA linkers (Kagalwala et al. 2004; Schwanbeck et al. 2004). These studies are consistent with earlier findings demonstrating the importance of the DNA linker both for ISWI binding and for subsequent ATPase and remodeling activities. Further, the internal nucleosome binding site of ISWI is in close proximity to the region where the H4 tails emit from the nucleosome, supporting the requirement of residues 16–19 of the histone H4 tail for ISWI activity (Clapier et al. 2001).

Additional studies with the SWI/SNF-family remodeler, RSC, extend these observations and provide an interesting comparison to ISWI. In contrast to

ISWI, RSC binds and remodels the nucleosome without a requirement for a DNA linker (Saha et al. 2005). Here, RSC appears to bind the nucleosome in either of two symmetrically equivalent orientations. Importantly, the ATPase domain of the RSC catalytic subunit, Sth1, binds nucleosomal DNA at a fixed internal site about two turns from the nucleosomal dyad. Although the studies with ISWI did not directly test whether the ATPase domain itself engages the internal site, both studies are consistent with the ATPase domain of remodelers engaging the nucleosome at an internal site about two turns from the nucleosomal dyad. Notably, interaction of the Sth1 ATPase domain with an unmodified mononucleosome ($K_d \sim 100$ nM) is weaker than with the intact RSC complex ($K_d \sim 10$ nM), suggesting that other components of RSC confer high-affinity binding. This is further supported by recent findings suggesting that SWI/SNF remodelers bind the nucleosome in a large pocket (Leschziner et al. 2005; Saha et al. 2005). These studies collectively indicate that the translocase domain of remodelers anchors to the nucleosome at an internal site.

Recent evidence also suggests that the ATPase domain of RSC conducts directional DNA translocation from this fixed internal site, drawing in DNA from the proximal linker and pumping it towards the dyad. Experimental evidence involved analysis of nucleosomes containing DNA linkers of varying length on one side of the nucleosome, as well as a series of nucleosomes each containing a gap in one strand of the DNA that were placed at different translational positions along the nucleosomal DNA which prevent translocation when the gap is encountered. Interestingly, the length of intact DNA present on one side of the nucleosome determined the length of DNA emitted from the opposite side of the nucleosome, with all substrate–product relationships consistent with directional DNA translocation initiating from, and terminating at, an internal position located two turns from the dyad. This interpretation is, in retrospect, consistent with earlier studies with RSC or SWI/SNF that identified remodeling products where the DNA is recessed up to 50 bp within the nucleosome disrupting up to five histone–DNA contacts (Flaus and Owen-Hughes 2003b; Kassabov et al. 2003).

Translocation along the DNA duplex also involves rotation along the helical DNA backbone and can result in the accumulation of twist if the DNA is constrained (Janscak and Bickle 2000). Accordingly, if remodelers conduct translocation from a fixed internal site, while attached to the histone octamer, then the rotation of DNA is constrained and DNA twist will accumulate (Havas et al. 2000). Thus, twist generation is a property consistent with DNA translocation.

4.2

Helicases/Translocases Provide Models for DNA Translocation by Remodelers

Much of our understanding of the mechanism of DNA translocation by monomeric translocases is based on the crystal structures of PcrA, NS3,

RecG, and Rad54 (Kim et al. 1998; Singleton et al. 2001; Velankar et al. 1999; Durr et al. 2005). Here, we will discuss these studies and in the next section apply these principles to build models for remodeler function. Crystal structures of these enzymes suggest that the translocation domain can be divided into two subdomains, termed the torsion and tracking domains. In the absence of nucleotide, both subdomains interact with DNA, and between them is a small cleft containing the intervening DNA. The torsion domain interacts with the duplex ahead of the tracking domain and, upon nucleotide binding, pulls and twists the DNA duplex, placing an additional 1 bp of DNA in the cleft between the two domains, as in PcrA (Dillingham et al. 2000; Velankar et al. 1999). The translocase domain includes two tandemly arranged RecA-like motifs, between which lies a pocket for nucleotide binding as well as a platform for DNA interaction. Nucleotide hydrolysis induces a conformational change between the two RecA-like domains that allows the enzyme to track along the DNA 1 bp in the $3'$–$5'$ direction. Put simply, the torsion domain feeds 1 bp of DNA to the tracking domain, which then ratchets forward one base, followed by the resetting of the torsion domain 1 bp forward. Thus, for PcrA the structural and biochemical evidence supports a step size of 1 bp/ATP hydrolyzed (Dillingham et al. 2000; Velankar et al. 1999).

4.3
Applying Principles of Translocases to Remodel Nucleosomes

We now apply the principles discussed above to form speculative models regarding the mechanism for SWI/SNF and ISWI family remodelers. Here, we emphasize that not all members of these families have been tested for their remodeling and DNA translocation properties, and therefore these models extrapolate from existing data on the members that have been tested. Furthermore, certain features may be clear for one remodeler family, but not directly tested for the other. Thus, these models remain speculative and are discussed to provide a framework for further testing and to stimulate discussion.

We suggest that SWI/SNF and ISWI share four steps in the remodeling process, but regulate these steps differently. They include: engagement of the nucleosome at a fixed position, a nucleotide-dependent conformational change in the remodeler that affects histone–DNA interactions, the ATP-dependent directional translocation of DNA from an internal site, and the propagation of a DNA wave around the nucleosome by one-dimensional diffusion.

First, the remodeler binds the nucleosome core with the ATPase domain engaging the DNA about two turns from the dyad (Fig. 3, step 1). The other proteins in the remodeler likely contribute to high-affinity binding and may also recognize histone modifications to help select particular nucleosomes for remodeling. Furthermore, the C-terminus of the ISWI ATPase subunit, but not SWI/SNF, makes a second contact with the DNA near the entry/exit site and the linker, which may also serve to regulate ISWI binding and its ATPase

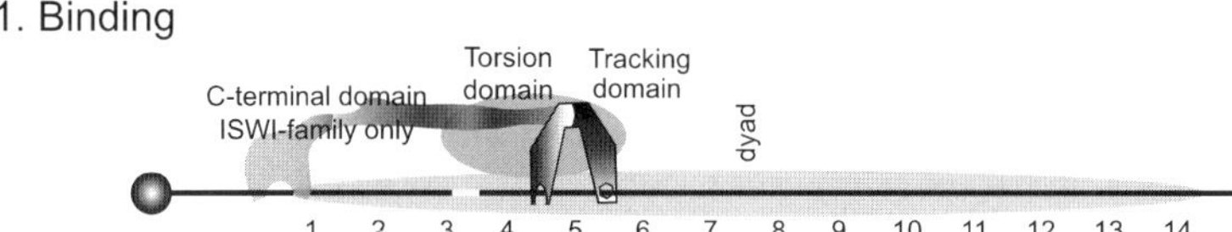

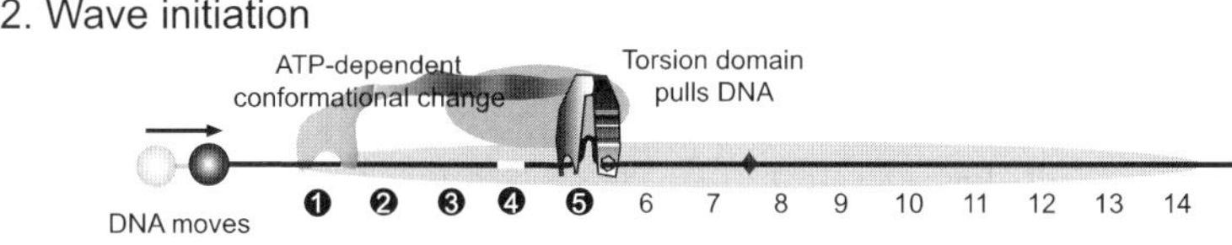

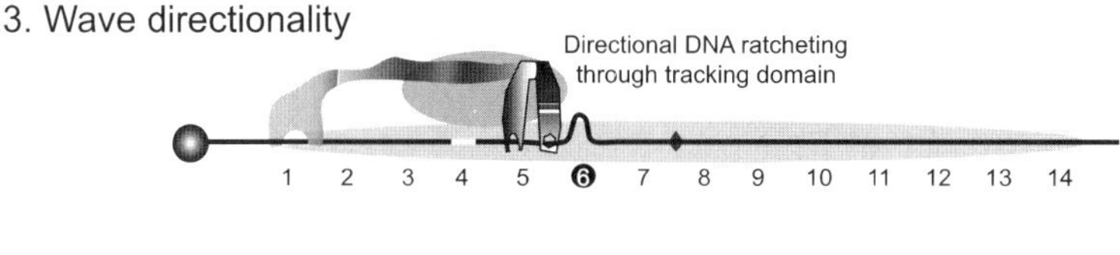

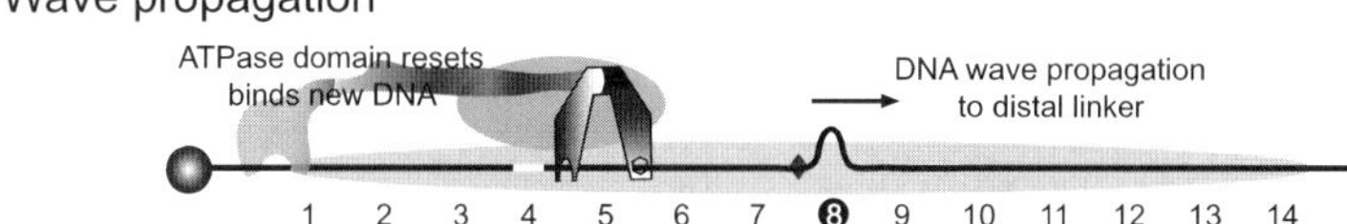

Fig. 3 Model for DNA translocation on a nucleosome. Nucleosomes are depicted in 2D with the histone octamer position depicted by the *grey oval*. DNA is indicated as a *black line*, with a segment near the translocase in *white* to illustrate translocation. Numbers 1 to 14 indicate histone–DNA contacts, either intact or broken (denoted by *black circle*). The remodeler ATPase subunit is divided into torsion and tracking domains. Also, as depicted the ISWI remodeler (and not SWI/SNF) contains an additional C-terminal domain

activity (Grune et al. 2003). Thus, both remodelers engage the nucleosome in a fixed manner, but they select substrates differently.

In the absence of nucleotide, the torsion and tracking domains are in an extended open conformation. ATP binding causes a conformational change between these domains, resulting in the closing of the cleft between them. This results in the torsion domain pulling DNA from the linker and placing the additional DNA in the cleft between the two domains (Fig. 3, step 2). The length of DNA drawn into the nucleosome is dependent on the translocation properties of the remodeler and will be discussed later. Consistent with this, studies with ISWI remodelers have shown that ATP binding causes a conformational change in the protein and also alters the interaction of ISWI with DNA (Fitzgerald et al. 2004).

At this point, a strained segment of DNA (DNA wave) resides in the cleft between the torsion and tracking domains. The tracking domain uses one strand of the DNA duplex for tracking and, in concert with ATP hydrolysis, allows this strained segment of DNA to pass through it, in one direction, 3′ to 5′ with respect to the tracking strand, thus serving as a molecular ratchet (Fig. 3, step 3). This places the undertwisted DNA wave near the nucleosomal dyad, which then propagates around the nucleosome by one-dimensional diffusion, breaking histone–DNA contacts at the leading edge and replacing them at the lagging edge (Fig. 3, step 4). Resolution of the DNA wave in the distal linker results in the sliding of the nucleosome by a distance equivalent to the length of the step size. ATP hydrolysis or ADP/Pi release resets the torsion domain to its original conformation, re-establishing the extended form of the cleft for another round of translocation. We term this model wave–ratchet–wave: a DNA wave is generated by torsion, passed through a directional ratchet, and then propagated to the distal linker by diffusion.

The length of DNA translocated in each round of translocation is dependent on the step size of the translocation, which is not known for any remodeler. However, the tracking requirement of 1 bp for RSC provides a minimal estimate, and is consistent with the 1 bp step size of PcrA (Dillingham et al. 2000; Saha et al. 2005). If the step size is 1 bp, then the histone–DNA contact proximal to the torsion domain will be strained and undertwisted by 1 bp. This strain could be eventually resolved by iteratively breaking and reforming contacts and transferring the strain to the linker region, resulting in the pulling of 1 bp DNA from the linker region. Alternatively, if the step size is more than 1 bp, then all four contacts may be simultaneously and transiently broken to provide sufficient DNA. This disruption would be followed by rapid reformation of energetically favored histone–DNA contacts at a new translational position. One interpretation consistent with this observation is that the conformational change in ISWI involves the concerted lifting of the DNA from the octamer near the entry/exit site (Strohner et al. 2005).

The size of the DNA wave is minimally equal to the step size of the translocation; however, it is possible that multiple rounds of translocation are required to generate a bigger wave of sufficient energy to break histone–DNA contacts near the dyad, which are energetically stronger than the contacts near the nucleosome entry/exit sites (Brower-Toland et al. 2002). Binding of the ATPase domain at a position about two turns from the dyad may assist remodelers in the efficient disruption of these stronger histone–DNA contacts. Furthermore, the placement of the tracking domain near the dyad allows the ratchet to ensure that wave movement is unidirectional and resolves into the opposite linker to provide a productive round of translocation.

Interaction between the ISWI C-terminal domain and DNA located at the entry/exit site of the nucleosome appears to regulate DNA translocation by ISWI; linker DNA is required for remodeling and DNA will not be translocated beyond the entry/exit with ISWI remodelers. In sharp contrast, SWI/SNF remodelers do not contact linker DNA and continue to translocate DNA until it reaches the translocase domain, ~ 50 bp inside the nucleosome. Thus, both remodelers may undergo a nucleotide-dependent conformational change to initiate disruption of histone–DNA contacts near the entry/exit, but only ISWI activity is regulated by the presence of linker DNA, a regulation likely important for its biological tasks.

Currently, it remains unclear whether remodelers render nucleosomal DNA accessible by sliding the DNA into the linker region or by transient access of the DNA on the surface of the nucleosome. At present, certain studies favor access on the surface, while others point to access via sliding into the linker (Narlikar et al. 2001; Saha et al. 2005). We note that these two modes of access are not mutually exclusive; remodelers may enable access by both modes, though understanding the primary mode has both mechanistic and biological implications. For access to occur on the surface, a segment of DNA has to be exposed. In principle, this could occur by peeling of the DNA from the edge of the nucleosome, through a change in the conformation or composition of the octamer, or in keeping with the DNA translocation model via the generation of a DNA wave. For accessibility on the surface of the nucleosome, the wave size is an important parameter, as it must be of sufficient size to enable factor access. Also, the dwell time of the wave will determine accessibility, since rapidly propagating DNA waves provide little or no opportunity for factor binding. Here, the remodeler may have the capacity to restrict wave propagation and resolution in the linker, allowing the wave size to increase through several rounds of ATP hydrolysis. Testing these properties will present challenges that may require single-molecule approaches.

4.4
DNA Translocation May Underlie DNA Twisting

Remodelers can generate superhelical torsion, as monitored by the ability of remodelers to generate a cruciform structure on an inverted repeat sequence $[AT]_n$, under negative superhelical tension (Havas et al. 2000). Here, the SWI/SNF complex, its catalytic subunit, Brg1, and ISWI generated superhelical torsion. Interestingly, SWI/SNF and its catalytic subunit generated torsion on both DNA and chromatin templates, whereas ISWI was only functional on a chromatin template, consistent with the requirement of the octamer by ISWI for full activity. Generation of superhelical torsion has raised the interesting possibility that remodelers work primarily by twisting DNA, which might serve to break histone–DNA contacts (Flaus and Owen-Hughes 2003a). Interestingly, the nucleosome can contain overtwisted segments of DNA between the histone–DNA contacts resulting in a twist defect (Luger et al. 1997; Richmond and Davey 2003; Suto et al. 2003). Thus, it has been proposed that a segment of DNA bearing a twist defect within the nucleosomal DNA would propagate around the histone octamer, resulting in the sliding of the DNA in 1 bp increments in a 'cork-screw'-like manner (Flaus and Owen-Hughes 2003a; Suto et al. 2003). Hence, remodelers might accelerate the rate of twist diffusion with minimal disruption of the histone–DNA contacts. For nucleosome mobilization by twist diffusion, a twisted segment would have to propagate through each histone–DNA contact sequentially around the entire length of the nucleosome.

To address whether twist diffusion is required for remodeling, several studies tested nucleosomes with DNA alterations such as single-stranded gaps of varying length, nicks, abasic sites, or steric blocks that should compromise or eliminate the ability to propagate twist, and found modest or no impact on remodeling efficiency (Langst and Becker 2001; Lorch et al. 2005; Saha et al. 2002, 2005). In one study, five base gaps placed in one strand of the DNA at multiple positions on the nucleosome that should all equally compromise twist affected restriction enzyme access in a position-dependent manner compared to that of intact DNA (Saha et al. 2005). An additional study attached a paramagnetic bead to the nucleosomal DNA, which should sterically prevent twist, and observed little affect on remodeling by ISWI complexes (Strohner et al. 2005).

Taken together, twist diffusion is apparently not required for remodeling in vitro. However, twist diffusion is an aspect of the directional DNA translocation model since translocation has both translational and twist components, with the twist component generated by the rotation of the helical DNA as it translocates through the tracking domain. We suggest that twist diffusion does occur during remodeling, but that translational movement of the DNA by the remodeler is the critical aspect, with twist diffusion utilized to help the DNA wave propagate efficiently.

5
Chromatin Remodeling Enables Specialized Biological Functions

Both SWI/SNF and ISWI remodelers are capable of altering chromatin, but with marked differences. This allows remodelers to perform specialized functions that are unique to either family. Here we discuss some of these unique functions.

5.1
Nucleosome Assembly and Spacing

The assembly and spacing of periodic nucleosome arrays is an ATP-dependent process catalyzed by the ISWI family of remodelers. Here, the C-terminus of the ISWI ATPase might restrict DNA translocation beyond the nucleosomal boundary by interacting with the linker DNA near the entry/exit site. Hence, translocation results in movement of nucleosomes to either end, resulting in the generation of predominantly end-positioned nucleosomes (Kagalwala et al. 2004). In contrast, in ISWI-containing complexes, the additional subunits might interact with distal linker DNA and alter its translocation limits. In fact, in the ISW2 complex, the Itc1 subunit interacts with a more distal region of linker DNA and results in the generation of predominantly centrally positioned nucleosomes (Kagalwala et al. 2004). One possibility is that DNA translocation will cease when the additional subunit can no longer bind DNA, thus restricting the translocation limit. Furthermore, non-ATPase subunits might sterically block the sliding of the nucleosome in an array, when they encounter another nucleosome, resulting in the generation of regularly spaced arrays. Here, it is predicted that the spacing distance will vary and depend on the additional subunits of the ISWI complex. In addition, ISWI complexes like ACF can generate periodically assembled nucleosomes in the presence of ATP and the histone chaperone Nap1 (Ito et al. 1997). This assembly process displays template commitment, which is likely due to DNA translocation (Fyodorov and Kadonaga 2002). Here, one possible way of assembling periodic nucleosomes is by the coupling of ATP-dependent spacing activity of the ISWI remodeler ACF to the assembly by histone chaperone Nap1.

5.2
Histone Octamer Transfer

SWI/SNF remodelers are able to transfer the histone octamer from a nucleosome to a naked DNA in *trans* (Lorch et al. 1999; Phelan et al. 2000). This *trans* displacement suggests that the SWI/SNF remodelers are able to disrupt nucleosome structure in a way that is distinct from the ISWI remodelers. This specialized property of the SWI/SNF remodelers might be explained by DNA translocation ~ 50 bp beyond the nucleosomal boundary which would result in nucleosomes lacking approximately five histone–DNA contacts from the entry/exit. Here, an acceptor DNA might bind to the exposed histone–DNA contacts generated by sliding of the nucleosomal DNA. Subsequent invasion of the free DNA by translocation can result in the transfer of histone octamers from one DNA template to another. This specialized remodeling property might have in vivo implications during gene activation, when nucleosomes are depleted from active regulatory elements (Lee et al. 2004).

5.3
Nucleosome Ejection

Recent studies of the yeast *PHO5* gene suggest that nucleosomes are removed from the promoter upon activation (Boeger et al. 2004; Reinke and Horz 2003). This occurs by complete disassembly as shown using topological analysis of chromatin circles formed from the activated promoter (Boeger et al. 2004). Interestingly, genome-wide analysis indicates that nucleosome loss is an attribute of highly active genes (Lee et al. 2004). These results suggest that active ejection of nucleosomes results in chromatin accessibility. Besides gene activation, removal of histone proteins also provides an opportunity to remove covalent marks on nucleosomes. Possible mechanisms of remodelers in facilitating nucleosome disassembly are yet to be determined. Interestingly, SWI/SNF remodelers facilitate ejection and/or octamer transfer in vitro (Lorch et al. 1999; Phelan et al. 2000). Ejection might also involve the chromatin assembly factor Asf1, since *asf1*Δ mutants are defective in both nucleosome ejection and PHO5 activation (Adkins et al. 2004). Asf1 could possibly function as a histone octamer acceptor protein during ejection as SWI/SNF and Asf1 display both genetic and physical interactions in *Drosophila* (Moshkin et al. 2002).

6
Conclusion

Over the last few years, remarkable progress has been made in understanding the mechanism of chromatin remodeling. Remodelers are now emerging as sophisticated molecular machines that are specialized to select and be regulated by particular nucleosome substrates based on features of the linker and histone modification state. Unique proteins in each remodeler complex mediate this specialization and tailor it for specific biological tasks including chromatin assembly, nucleosome ejection, and DNA repair. Here, we have reviewed recent evidence that SWI/SNF- and ISWI-family remodelers utilize DNA translocation as an aspect of their mechanism, and provided a model to stimulate discussion about how DNA translocation might be applied to nucleosomes. The future for studying remodelers will include additional genetic and biochemical effort to understand how unique components specialize remodelers and select substrates, genome-wide localization to understand their sites of action in vivo, single-molecule analyses, and structural approaches to probe the remodeler mechanism. Only through these combined approaches will the true picture of their functions emerge.

Acknowledgements Because of space restrictions, we apologize to those whose work could not be discussed. We thank Maggie Kasten for comments.

References

Adkins MW, Howar SR, Tyler JK (2004) Chromatin disassembly mediated by the histone chaperone Asf1 is essential for transcriptional activation of the yeast PHO5 and PHO8 genes. Mol Cell 14(5):657–666

Alexeev A, Mazin A, Kowalczykowski SC (2003) Rad54 protein possesses chromatin-remodeling activity stimulated by the Rad51-ssDNA nucleoprotein filament. Nat Struct Biol 10(3):182–186

Almer A et al. (1986) Removal of positioned nucleosomes from the yeast PHO5 promoter upon PHO5 induction releases additional upstream activating DNA elements. Embo J 5(10):2689–2696

Becker PB, Horz W (2002) ATP-dependent nucleosome remodeling. Annu Rev Biochem 71:247–273

Boeger H et al. (2004) Removal of promoter nucleosomes by disassembly rather than sliding in vivo. Mol Cell 14(5):667–673

Bowen NJ et al. (2004) Mi-2/NuRD: multiple complexes for many purposes. Biochim Biophys Acta 1677(1–3):52–57

Brehm A et al. (2000) dMi-2 and ISWI chromatin remodelling factors have distinct nucleosome binding and mobilization properties. Embo J 19(16):4332–4341

Brower-Toland BD et al. (2002) Mechanical disruption of individual nucleosomes reveals a reversible multistage release of DNA. Proc Natl Acad Sci USA 99(4):1960–1965

Cairns BR (1998) Chromatin remodeling machines: similar motors, ulterior motives. Trends Biochem Sci 23(1):20–25

Cairns BR (2005) Chromatin remodeling complexes: strength in diversity, precision through specialization. Curr Opin Genet Dev 15(2):185–190

Cairns BR et al. (1994) A multisubunit complex containing the SWI1/ADR6, SWI2/SNF2, SWI3, SNF5, and SNF6 gene products isolated from yeast. Proc Natl Acad Sci USA 91(5):1950–1954

Cairns BR et al. (1996) RSC, an essential, abundant chromatin-remodeling complex. Cell 87(7):1249–1260

Chai B et al. (2005) Distinct roles for the RSC and Swi/Snf ATP-dependent chromatin remodelers in DNA double-strand break repair. Genes Dev 19(14):1656–1661

Clapier CR et al. (2001) Critical role for the histone H4 N terminus in nucleosome remodeling by ISWI. Mol Cell Biol 21(3):875–883

Corey LL et al. (2003) Localized recruitment of a chromatin-remodeling activity by an activator in vivo drives transcriptional elongation. Genes Dev 17(11):1392–1401

Corona DF, Tamkun JW (2004) Multiple roles for ISWI in transcription, chromosome organization and DNA replication. Biochim Biophys Acta 1677(1–3):113–119

Corona DF et al. (2002) Modulation of ISWI function by site-specific histone acetylation. EMBO Rep 3(3):242–247

Cote J, Peterson CL, Workman JL (1998) Perturbation of nucleosome core structure by the SWI/SNF complex persists after its detachment, enhancing subsequent transcription factor binding. Proc Natl Acad Sci USA 95(9):4947–4952

Cote J et al. (1994) Stimulation of GAL4 derivative binding to nucleosomal DNA by the yeast SWI/SNF complex. Science 265(5168):53–60

Dillingham MS, Wigley DB, Webb MR (2000) Demonstration of unidirectional single-stranded DNA translocation by PcrA helicase: measurement of step size and translocation speed. Biochemistry 39(1):205–212

Durr H et al. (2005) X-ray structures of the Sulfolobus solfataricus SWI2/SNF2 ATPase core and its complex with DNA. Cell 121(3):363–373

Eisen JA, Sweder KS, Hanawalt PC (1995) Evolution of the SNF2 family of proteins: subfamilies with distinct sequences and functions. Nucleic Acids Res 23(14):2715–2723

Fan HY et al. (2003) Distinct strategies to make nucleosomal DNA accessible. Mol Cell 11(5):1311–1322

Fitzgerald DJ et al. (2004) Reaction cycle of the yeast Isw2 chromatin remodeling complex. Embo J 23(19):3836–3843

Flaus A, Owen-Hughes T (2003a) Mechanisms for nucleosome mobilization. Biopolymers 68(4):563–578

Flaus A, Owen-Hughes T (2003b) Dynamic properties of nucleosomes during thermal and ATP-driven mobilization. Mol Cell Biol 23(21):7767–7779

Fyodorov DV, Kadonaga JT (2002) Dynamics of ATP-dependent chromatin assembly by ACF. Nature 418(6900):897–900

Gottesfeld JM, Luger K (2001) Energetics and affinity of the histone octamer for defined DNA sequences. Biochemistry 40(37):10927–10933

Grune T et al. (2003) Crystal structure and functional analysis of a nucleosome recognition module of the remodeling factor ISWI. Mol Cell 12(2):449–460

Guyon JR et al. (1999) Stable remodeling of tailless nucleosomes by the human SWI-SNF complex. Mol Cell Biol 19(3):2088–2097

Hamiche A et al. (1999) ATP-dependent histone octamer sliding mediated by the chromatin remodeling complex NURF. Cell 97(7):833–842

Havas K et al. (2000) Generation of superhelical torsion by ATP-dependent chromatin remodeling activities. Cell 103(7):1133–1142

Henikoff S, Furuyama T, Ahmad K (2004) Histone variants, nucleosome assembly and epigenetic inheritance. Trends Genet 20(7):320–326

Huang J, Hsu JM, Laurent BC (2004) The RSC nucleosome-remodeling complex is required for Cohesin's association with chromosome arms. Mol Cell 13(5):739–750

Imbalzano AN, Schnitzler GR, Kingston RE (1996) Nucleosome disruption by human SWI/SNF is maintained in the absence of continued ATP hydrolysis. J Biol Chem 271(34):20726–20733

Ito T et al. (1997) ACF, an ISWI-containing and ATP-utilizing chromatin assembly and remodeling factor. Cell 90(1):145–155

Janscak P, Bickle TA (2000) DNA supercoiling during ATP-dependent DNA translocation by the type I restriction enzyme EcoAI. J Mol Biol 295(4):1089–1099

Jaskelioff M et al. (2003) Rad54p is a chromatin remodeling enzyme required for heteroduplex DNA joint formation with chromatin. J Biol Chem 278(11):9212–9218

Jenuwein T, Allis CD (2001) Translating the histone code. Science 293(5532):1074–1080

Kagalwala MN et al. (2004) Topography of the ISW2-nucleosome complex: insights into nucleosome spacing and chromatin remodeling. Embo J 23(10):2092–2104

Kassabov SR et al. (2003) SWI/SNF unwraps, slides, and rewraps the nucleosome. Mol Cell 11(2):391–403

Kim JL et al. (1998) Hepatitis C virus NS3 RNA helicase domain with a bound oligonucleotide: the crystal structure provides insights into the mode of unwinding. Structure 6(1):89–100

Korber P, Horz W (2004) SWRred not shaken; mixing the histones. Cell 117(1):5–7

Kornberg RD, Lorch Y (1999) Twenty-five years of the nucleosome, fundamental particle of the eukaryote chromosome. Cell 98(3):285–294

Kwon H et al. (1994) Nucleosome disruption and enhancement of activator binding by a human SW1/SNF complex. Nature 370(6489):477–481

Langst G, Becker PB (2001) ISWI induces nucleosome sliding on nicked DNA. Mol Cell 8(5):1085–1092

Langst G et al. (1999) Nucleosome movement by CHRAC and ISWI without disruption or trans-displacement of the histone octamer. Cell 97(7):843–852

Laurent BC, Yang X, Carlson M (1992) An essential Saccharomyces cerevisiac gene homologous to SNF2 encodes a helicase-related protein in a new family. Mol Cell Biol 12(4):1893–1902

Lee CK et al. (2004) Evidence for nucleosome depletion at active regulatory regions genome-wide. Nat Genet 36(8):900–905

Leschziner AE et al. (2005) Structural studies of the human PBAF chromatin-remodeling complex. Structure (Camb) 13(2):267–275

Li G, Widom J (2004) Nucleosomes facilitate their own invasion. Nat Struct Mol Biol 11(8):763–769

Li G et al. (2005) Rapid spontaneous accessibility of nucleosomal DNA. Nat Struct Mol Biol 12(1):46–53

Lorch Y, Davis B, Kornberg RD (2005) Chromatin remodeling by DNA bending, not twisting. Proc Natl Acad Sci USA 102(5):1329–1332

Lorch Y, Zhang M, Kornberg RD (1999) Histone octamer transfer by a chromatin-remodeling complex. Cell 96(3):389–392

Lorch Y et al. (1998) Activated RSC-nucleosome complex and persistently altered form of the nucleosome. Cell 94(1):29–34

Luger K et al. (1997) Crystal structure of the nucleosome core particle at 2.8 A resolution. Nature 389(6648):251–260

Martens JA, Winston F (2003) Recent advances in understanding chromatin remodeling by Swi/Snf complexes. Curr Opin Genet Dev 13(2):136–142

Mellor J, Morillon A (2004) ISWI complexes in Saccharomyces cerevisiae. Biochim Biophys Acta 1677(1–3):100–112

Moshkin YM et al. (2002) Histone chaperone ASF1 cooperates with the Brahma chromatin-remodelling machinery. Genes Dev 16(20):2621–2626

Murray NE (2000) Type I restriction systems: sophisticated molecular machines. Microbiol Mol Biol Rev 64(2):412–434

Narlikar GJ, Fan HY, Kingston RE (2002) Cooperation between complexes that regulate chromatin structure and transcription. Cell 108(4):475–487

Narlikar GJ, Phelan ML, Kingston RE (2001) Generation and interconversion of multiple distinct nucleosomal states as a mechanism for catalyzing chromatin fluidity. Mol Cell 8(6):1219–1230

Owen-Hughes T (2003) Colworth memorial lecture. Pathways for remodelling chromatin. Biochem Soc Trans 31(Part 5):893–905

Peterson CL, Dingwall A, Scott MP (1994) Five SWI/SNF gene products are components of a large multisubunit complex required for transcriptional enhancement. Proc Natl Acad Sci USA 91(8):2905–2908

Phelan ML, Schnitzler GR, Kingston RE (2000) Octamer transfer and creation of stably remodeled nucleosomes by human SWI-SNF and its isolated ATPases. Mol Cell Biol 20(17):6380–6389

Reinke H, Horz W (2003) Histones are first hyperacetylated and then lose contact with the activated PHO5 promoter. Mol Cell 11(6):1599–1607

Richmond TJ, Davey CA (2003) The structure of DNA in the nucleosome core. Nature 423(6936):145–150

Saha A, Wittmeyer J, Cairns BR (2002) Chromatin remodeling by RSC involves ATP-dependent DNA translocation. Genes Dev 16(16):2120–2134

Saha A, Wittmeyer J, Cairns BR (2005) Chromatin remodeling through directional DNA translocation from an internal nucleosomal site. Nat Struct Mol Biol 12(9):747–755

Schnitzler G, Sif S, Kingston RE (1998) Human SWI/SNF interconverts a nucleosome between its base state and a stable remodeled state. Cell 94(1):17–27

Schwanbeck R, Xiao H, Wu C (2004) Spatial contacts and nucleosome step movements induced by the NURF chromatin remodeling complex. J Biol Chem 279(38):39933–39941

Shen X et al. (2000) A chromatin remodelling complex involved in transcription and DNA processing. Nature 406(6795):541–544

Singleton MR, Wigley DB (2002) Modularity and specialization in superfamily 1 and 2 helicases. J Bacteriol 184(7):1819–1826

Singleton MR, Scaife S, Wigley DB (2001) Structural analysis of DNA replication fork reversal by RecG. Cell 107(1):79–89

Soultanas P et al. (2000) Uncoupling DNA translocation and helicase activity in PcrA: direct evidence for an active mechanism. Embo J (14):3799–3810

Strohner R et al. (2005) A 'loop recapture' mechanism for ACF-dependent nucleosome remodeling. Nat Struct Mol Biol 12(8):683–690

Suto RK et al. (2003) Crystal structures of nucleosome core particles in complex with minor groove DNA-binding ligands. J Mol Biol 326(2):371–380

Thoma NH et al. (2005) Structure of the SWI2/SNF2 chromatin-remodeling domain of eukaryotic Rad54. Nat Struct Mol Biol 12(4):350–356

Tsukiyama T, Wu C (1995) Purification and properties of an ATP-dependent nucleosome remodeling factor. Cell 83(6):1011–1020

van Attikum H, Gasser SM (2005) The histone code at DNA breaks: a guide to repair? Nat Rev Mol Cell Biol 6(10):757–765

Varga-Weisz PD et al. (1997) Chromatin-remodelling factor CHRAC contains the ATPases ISWI and topoisomerase II. Nature 388(6642):598–602

Velankar SS et al. (1999) Crystal structures of complexes of PcrA DNA helicase with a DNA substrate indicate an inchworm mechanism. Cell 97(1):75–84

Whitehouse I et al. (1999) Nucleosome mobilization catalysed by the yeast SWI/SNF complex. Nature 400(6746):784–787

Whitehouse I et al. (2003) Evidence for DNA translocation by the ISWI chromatin-remodeling enzyme. Mol Cell Biol 23(6):1935–1945

Widom J (1998) Structure, dynamics, and function of chromatin in vitro. Annu Rev Biophys Biomol Struct 27:285–327

Widom J (2001) Role of DNA sequence in nucleosome stability and dynamics. Q Rev Biophys 34(3):269–324

Winston F, Carlson M (1992) Yeast SNF/SWI transcriptional activators and the SPT/SIN chromatin connection. Trends Genet 8(11):387–391

Wu J, Grunstein M (2000) 25 years after the nucleosome model: chromatin modifications. Trends Biochem Sci 25(12):619–623

Subject Index

Printing: Krips bv, Meppel
Binding: Stürtz, Würzburg